2

# Experimental Studies of
# **Neutrino Oscillations**

# Experimental Studies of
# Neutrino Oscillations

## Takaaki Kajita

University of Tokyo, Kashiwa, Japan

 **World Scientific**

NEW JERSEY · LONDON · SINGAPORE · BEIJING · SHANGHAI · HONG KONG · TAIPEI · CHENNAI · TOKYO

*Published by*

World Scientific Publishing Co. Pte. Ltd.

5 Toh Tuck Link, Singapore 596224

*USA office:* 27 Warren Street, Suite 401-402, Hackensack, NJ 07601

*UK office:* 57 Shelton Street, Covent Garden, London WC2H 9HE

**British Library Cataloguing-in-Publication Data**
A catalogue record for this book is available from the British Library.

**EXPERIMENTAL STUDIES OF NEUTRINO OSCILLATIONS**

ISBN 978-981-4759-15-1
ISBN 978-981-4759-26-7 (pbk)

Printed by FuIsland Offset Printing (S) Pte Ltd Singapore

# Foreword

On Oct. 6, 2015, the Nobel Committee announced the award of the 2015 Physics Prize to Takaaki Kajita (Super-Kamiokande Collaboration) and Arthur B. McDonald (Sudbury Neutrino Observatory) in honor of their key contributions to our understanding of the nature and metamorphosis of the neutrino.

This volume of collected works of Kajita on neutrino oscillations provides a good glimpse into as well as a record of the rise and the role of Asian research in the frontiers of neutrino physics.

The neutrino was proposed by W. Pauli in a postcard dated Dec. 4, 1930 to Lise Meitner *et al.* (Radioactive Ladies and Gentlemen), and Pauli gave a formal talk on this idea at the Solvay Congress 1933, Oct. 22–29. Fermi was at the Congress, and on Dec. 31, 1933 he published in Ricerca Scientifica the first exposition on the 4-fermion theory of weak interaction. It was he who gave the name "little neutral one" to this invention by Pauli. Interestingly, this first paper had been rejected by Nature.

Efforts to detect the mysterious little neutral one were not successful. The elusive neutrino was finally observed a quarter of a century later by Reines and Cowan in 1956 at the Savannah River reactor facility in South Carolina. They used two tanks with 200 liters of water.

This dwarfs in comparison to the size of detectors in the ones employed at Kamiokande (a giant tank containing 3,000 tons of pure water and 1,000 photomultiplier tubes, PMT) and even more outsized at Super-Kamiokande, with 50,000 tons of water and 13,000 PMT.

One of the first people Kajita called after receiving news of the Nobel Prize was the 2002 Nobel Prize winner Masatoshi Koshiba, his mentor and fellow neutrino researcher. It was Koshiba's insight and foresight to recognize how Japan could seize an opportunity to be ahead of the IMB (Irvine Michigan Brookhaven) effort in the race to look for nucleon decay. While IMB uses 8,000 tons of water with 2048 PMT, Koshiba's strategy was not to revise his proposal to ask the government to build bigger, but to build better. Where the IMB used PMT of 12.5 cm (later upgraded to PMTs of 20 cm) in diameter, he convinced Hamamatsu to develop a 50 cm PMT that enhanced the sensitivity by orders of magnitude.

When Kamiokande succeeded in detecting the neutrinos from the 1987A Supernova, the status of Japan was established in the world of physics.

This was quickly followed by the first measurements of the deficit of $\nu_\mu$ vs. $\nu_e$ in the atmospheric neutrinos, the first indication of neutrino oscillation. Kamiokande was quickly upgraded to Super-Kamiokande I (SK-I), and the long baseline neutrino oscillation experiment K2K (from KEK to Kamioka) was started. Super-K was successively upgraded to SK-II, SK-III, SK-IV.

As the talks included in this volume show, they resulted in great strides in the study of solar neutrino, atmospheric neutrino, and in the study of neutrino oscillations through T2K (Tokai to Kamioka).

Japan is now a major force in the study of the 3 families of neutrinos. Much remains to be done to clarify the Dirac vs. Majorana nature of the neutrino, and the cosmological implications of the neutrino.

The collected works of Kajita and his Super-Kamiokande group will leave an indelible foot-print in the history of big and better science.

Ngee-Pong Chang
*November, 2015*

# Contents

# Chapter 1

# Neutrino Oscillations: Discovery, Current Status, Future Directions[*]

Takaaki Kajita

*ICRR and IPMU, University of Tokyo, Kashiwa, Chiba 277-8582, Japan*
*kajita@icrr.u-tokyo.ac.jp*

NEUTRINO oscillation was discovered about 10 years ago. Since then, the knowledge on neutrino masses and mixing angles have been improving substantially. This article describes neutrino oscillation experiments; the discovery, the present status and the future prospect.

*Keywords*: Neutrino; neutrino oscillation.

## 1. Introduction

In 1933, at the Solvay conference, W. Pauli presented his idea on neutrinos in public.[1] 75 years later, it is understood that neutrinos are still very important for the deeper understanding of elementary particle physics. Small but finite neutrino masses are believed to be related to physics at a very high energy scale.[2–4] The observed large neutrino mixing angles might be the hint for understanding physics at the very high energies. Furthermore, the physics of neutrino masses might be related to the baryon asymmetry of the

---

[*]This article was originally published in *Int. J. Mod. Phys. A* **24**, 3437 (2009).

Universe.[5] As of this writing, neutrino oscillation experiments give one of a few experimental evidences for "physics beyond the standard model".

One of the most sensitive methods to observe small neutrino masses is to study neutrino flavor oscillations.[6,7] If neutrinos have finite masses, each flavor eigenstate (for example, $\nu_\mu$) can be expressed by a combination of mass eigenstates ($\nu_1$, $\nu_2$ and $\nu_3$). For simplicity let us discuss two flavor neutrino oscillations. The probability for a neutrino produced in a flavor state $\nu_\alpha$ to be observed in a flavor state $\nu_\beta$ after traveling a distance $L$ through the vacuum is:

$$P(\nu_\alpha \rightarrow \nu_\beta) = \sin^2 2\theta_{ij} \sin^2 \left( \frac{1.27 \Delta m_{ij}^2 (\text{eV}^2) L(\text{km})}{E_\nu(\text{GeV})} \right), \qquad (1)$$

where $E_\nu$ is the neutrino energy, $\theta_{ij}$ is the mixing angle between the flavor eigenstates and the mass eigenstates, and $\Delta m_{ij}^2$ is the mass-squared difference of the neutrino mass eigenstates. The above description has to be generalized to three-flavor oscillations. However, it is known that it is approximately correct to assume two-flavor oscillations for analyses of the present neutrino oscillation data. Therefore, in this article, we mostly discuss two flavor neutrino oscillations.

## 2. Discovery of Neutrino Oscillations

In the early 1980's, several large underground detectors were constructed. The main motivation for such detectors was the discovery of proton decays. These experiments did not observe any evidence for proton decays. However, they observed a large number of atmospheric neutrino events, which are the most serious background for the proton decay search. Therefore, these experiments studied these atmospheric neutrino events.

In 1988, Kamiokande, a 4500 ton water Cherenkov detector, studied the number of $e$-like and $\mu$-like events, which were mostly CC $\nu_e$ and $\nu_\mu$ interactions, respectively. They found that the number of $\mu$-like events had a significant deficit compared with the Monte Carlo prediction, while the number of $e$-like events had in good agreement with the prediction.[8] The flavor ratio of the atmospheric neutrino flux, $(\nu_\mu + \overline{\nu}_\mu)/(\nu_e + \overline{\nu}_e)$, has been calculated accurately (better than 5%) in the relevant energy range. Because of the small $(\mu/e)_{Data}/(\mu/e)_{Prediction}$ ratio, it was concluded: "We are unable to explain the data as the result of the systematic detector effects or uncertainties in the atmospheric neutrino fluxes. Some as-yet-accounted-for physics such as neutrino oscillations might explain the data." This result was the beginning of the serious interest in atmospheric neutrinos. The subsequent study by the same experiment confirmed that the deficit of $\mu$-like events was explained well by either $\nu_\mu \rightarrow \nu_\tau$ or $\nu_\mu \rightarrow \nu_e$ oscillations.[9] A consistent result on $\nu_\mu$ deficit was reported in the early 1990's from the IMB experiment.[10]

Another important hint toward the understanding of atmospheric neutrino phenomena was given in the mid. 1990's.[11] Zenith angle distributions for multi-GeV fully-contained events and partially contained events were studied in Kamiokande. For detectors near the surface of the Earth, the neutrino flight distance, and thus the neutrino oscillation probability, is a function of the zenith angle of the neutrino direction. Vertically downward-going neutrinos travel about 15 km while vertically upward-going neutrinos travel about 12,800 km before interacting in the detector. The Kamiokande data showed that the deficit of $\mu$-like events depended on the neutrino zenith angle. However, due to the relatively poor event statistics, the statistical significance of the up-down asymmetry in the Kamiokande data was 2.9 standard deviations, and therefore the data were not conclusive.

It was obvious that a much larger detector was required to observe conclusive evidence for neutrino oscillations with high enough statistics. In 1996, the Super-Kamiokande experiment started. The detector is a 50 kton water Cherenkov detector with the fiducial mass of 22.5 kton. In 1998, the Super-Kamiokande

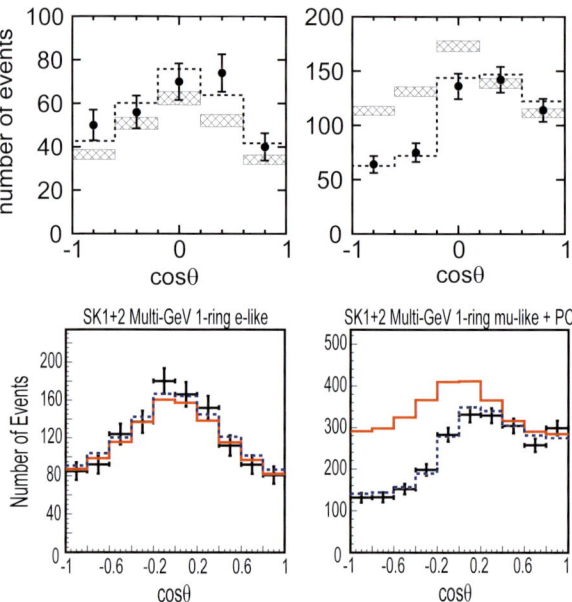

Figure 1.　Zenith angle distributions from Super-Kamiokande. The left and right panels show multi-GeV fully contained *e*-like events and multi-GeV fully contained *μ*-like plus partially-contained events, respectively. The top and bottom panels show the data as of 1998 and those of the full Super-K-I+II data, respectively.

experiment, with substantially larger data statistics than those in the previous experiments, concluded that the atmospheric neutrino data gave evidence for neutrinos oscillations.[12,13] The subsequent study confirmed that the oscillation was mostly between $\nu_\mu$ and $\nu_\tau$.[14]

Super-Kamiokande has been accumulating the data. Figure 1 shows the zenith angle distributions for multi-GeV fully-contained and partially-contained events as of 1998 (535 day exposure) and those of Super-K-I+II (1489+799 day exposure). It is clear that the data statistics have been improved substantially. Furthermore, evidence for the first oscillation minimum as expected by the sinusoidal oscillation probability has been observed as shown in Figure 2.[15] Figure 3 shows the allowed neutrino oscillation parameters from the Super-K atmospheric neutrino data together with

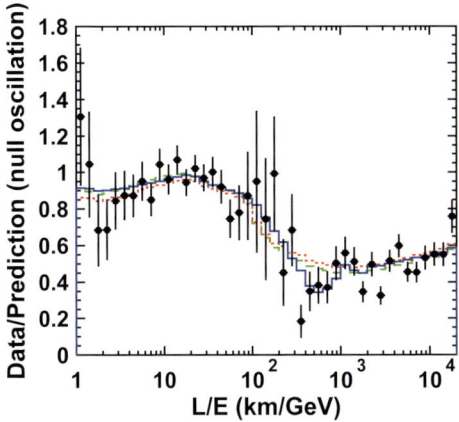

Figure 2. $L/E$ distribution observed in Super-Kamiokande. The plot is updated from Ref. 15. The solid (blue), dashed (green), and dotted (red) histograms show the best-fit predictions based on neutrino oscillation, decay and decoherence models, respectively.

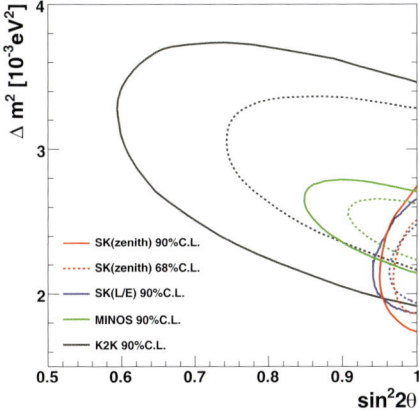

Figure 3. Allowed neutrino oscillation parameters from atmospheric neutrino data from Super-K and those from long baseline experiments (K2K[16] and MINOS[17]).

those from long baseline experiments. The oscillation parameters are understood accurately. Especially, the atmospheric neutrino experiments contribute to the measurement of the mixing parameter ($\sin^2 2\theta_{23}$).

## 3. Long Baseline Experiments

Probably, atmospheric neutrino experiments are suited to the discovery of oscillations, since atmospheric neutrinos have a wide neutrino flight length distribution, a wide neutrino energy coverage, and the beam is the mixture of $\nu_e$, $\bar{\nu}_e$, $\nu_\mu$ and $\bar{\nu}_\mu$. On the other hand, long baseline experiments have only one neutrino flight length, a controlled neutrino energy and an almost pure neutrino flavor. Therefore, in general, the long baseline experiments are suited to the precise measurements of the oscillation parameters.

The first long baseline neutrino oscillation experiment was the K2K experiment. The baseline length was 250 km between KEK and Super-K. The experiment started in 1999 and finished in 2004.[16] The MINOS experiment is the long baseline experiment between Fermilab and the Soudan mine. The baseline length is 735 km. The experiment started in 2005 and is still running. In both experiments, the deficits of $\nu_\mu$ events were clearly observed. Even more convincing results have been obtained from the measurements of the energy spectra from these experiments. For example, Figure 4 shows the neutrino energy spectrum measured by the MINOS experiment.[17]

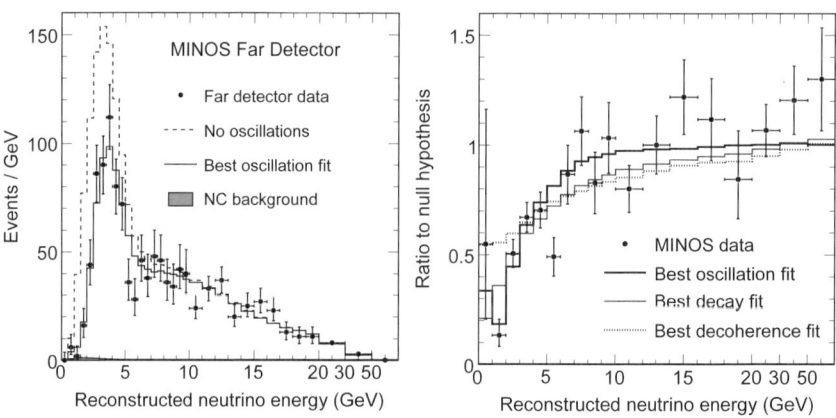

Figure 4.  Left: Neutrino energy spectrum measured by the MINOS experiment.  Right: Ratio of the data and the no-oscillation expectation. From Ref. 17.

The event rate in the 1 to 2 GeV energy range has a significant deficit compared with the no-oscillation expectation. It is clear that the deficit depends on neutrino energy and is completely consistent with the oscillation expectation.

## 4. Solar Neutrino Problem and Neutrino Oscillations

The measurement of solar neutrinos was initiated by Ray Davis Jr. and his collaborators in the 1960's at Homestake.[18] Soon after starting the measurement, they found that the observed flux was much lower than the prediction by the Standard Solar Model (SSM, see Ref. 19 for the recent one). Various possibilities, including neutrino oscillations in the vacuum, were discussed to explain the apparent inconsistency between the observation and the calculation.

In the mid 1980's, it was discovered that the effect of the solar matter might significantly enhance the neutrino oscillation probability even if the intrinsic mixing angle was small (MSW mechanism).[20,21] This mechanism naturally explained the solar neutrino problem, and therefore had a significant impact to the solar neutrino physics.

In 1989, Kamiokande observed the $^8B$ solar neutrino flux by neutrino electron scattering, $ve \rightarrow ve$.[22] This experiment confirmed that neutrinos are coming from the Sun and that the solar neutrino flux was significantly lower than the SSM prediction. Subsequently, solar neutrino experiments that used Gallium were carried out.[23,24] These experiments were unique in the sense that about half of the calculated event rate was due to fundamental $pp$ solar neutrinos. These low energy solar neutrino experiments also observed that the solar neutrino flux was lower than the SSM prediction. Although the solar neutrino deficit was clearly observed in these experiments, the cause of that deficit was not firmly identified, partly because no experiment observed the $v_e$ and $(v_\mu + v_\tau)$ fluxes separately. MSW mechanism was a serious possibility, but was not uniquely identified as the solution to this problem.

The situation changed dramatically in 2001 and 2002. In 2001, the SNO heavy water experiment measured the $^8B$ solar neutrinos by $v_e + D \rightarrow e^- + p + p$,[25] i.e., by charged current $v_e$ interactions. At that time Super-Kamiokande measured the $^8B$ neutrino flux precisely by neutrino-electron scattering.[26] By comparing these results, it was found that the flux measured by Super-Kamiokande (assuming all the neutrinos are electron-neutrinos) was significantly higher than that measured by SNO. This was a $3.3\sigma$ discrepancy, and was concluded as evidence for solar neutrino oscillations, since $v_\mu$'s and $v_\tau$'s generated by neutrino oscillations contribute only to the neutrino-electron scatterings.

In 2002, the evidence was substantially strengthened by the measurement of the total ($= v_e + v_\mu + v_\tau$) neutrino flux measurement in SNO. They observed $v_x + D \rightarrow v_x + p + n$.[27] The observed total flux was consistent with the SSM prediction, confirming the solar neutrino oscillation at the 5.5 standard deviation level.

Furthermore, a long baseline reactor neutrino experiment, Kam-LAND, observed not only the deficit[28] but also the spectrum distortion.[29] See Figure 5 for the most updated results from KamLAND.

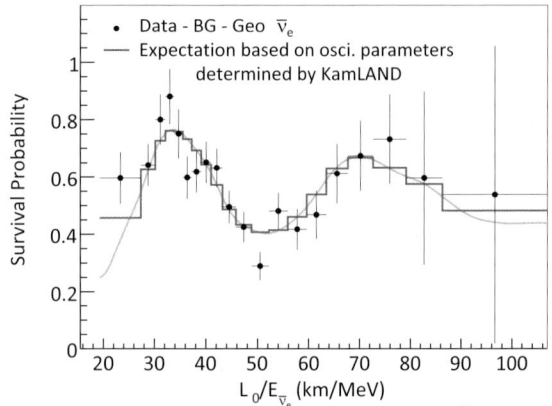

Figure 5.   Ratio of the background subtracted reactor $\bar{v}_e$ spectrum to the no-oscillation expectation as a function of $L_0/E$. 2881 ton·year of the KamLAND data are used. $L_0 = 180$ km is assumed. (From Ref. 38.)

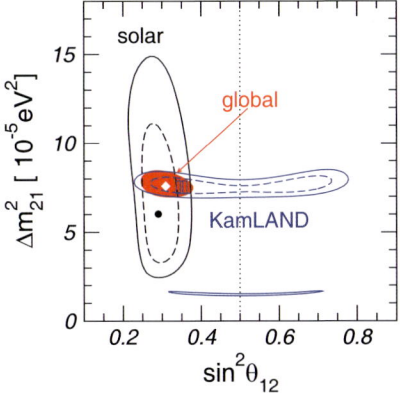

Figure 6.    Allowed region of neutrino oscillation parameters by the solar and the KamLAND experiments. (From Ref. 30.)

The observed spectrum distortion is regarded as very strong evidence for "oscillations". The mean flight length of the neutrinos detected by KamLAND was 180 km. The observed spectrum distortion together with the known flight length makes it possible to estimate the $\Delta m_{12}^2$ parameters precisely. Figure 6 shows the allowed $\Delta m_{12}^2$ and $\sin^2 \theta_{12}$ parameters. These parameters are already measured precisely.

## 5.  Future Directions

$\Delta m_{23}^2$, $\sin^2 2\theta_{23}$, $\Delta m_{12}^2$ and $\sin^2 \theta_{12}$ have been measured accurately by the present generation experiments assuming 2-flavor neutrino oscillations. However, we know that there are 3 neutrino flavors. Therefore, the neutrino oscillation studies should be generalized to 3-flavor oscillations. We discuss the possible future direction of neutrino oscillation experiments.

### 5.1.  *Search for non-zero $\theta_{13}$*

For simplicity, we neglect the effect of the oscillations related to $\Delta m_{12}^2$. Under this approximation, there are only three oscillation parameters; $\theta_{13}$, $\theta_{23}$ and $\Delta m_{13}^2$ ($= \Delta m_{23}^2$). In this case, for example, the $\nu_e \to \nu_e$ and $\nu_\mu \to \nu_e$ oscillation probabilities in vacuum can be written as:

$$P(\nu_e \to \nu_e) = 1 - \sin^2 2\theta_{13} \sin^2 \left( \frac{1.27\Delta m_{13}^2 L}{E_\nu} \right), \qquad (2)$$

$$P(\nu_\mu \to \nu_e) = \sin^2 \theta_{23} \sin^2 2\theta_{13} \sin^2 \left( \frac{1.27\Delta m_{13}^2 L}{E_\nu} \right). \qquad (3)$$

The CHOOZ experiment searched for $\bar{\nu}_e$ disappearance in the reactor $\bar{\nu}_e$ flux based on Eq. 2. The observed flux was consistent with the no-oscillation expectation. The upper limit on $\sin^2 2\theta_{13}$ was $\sim 0.15$.[31]

In the near future, various experiments that are much more sensitive to a small $\theta_{13}$ than the present experiments will start taking data. They use either reactor neutrinos or accelerator neutrino beams. If a finite value of $\theta_{13}$ is observed, we understand the overall structure of the neutrino mixing matrix (except for the CP phase). Thus, $\theta_{13}$ should be measured with a high priority. Furthermore, the measurement of $\theta_{13}$ is very important for future neutrino oscillation experiments, which will be discussed in the next section.

Reactor experiments try to observe $\bar{\nu}_e$ disappearance (Eq. 2). At present 3 reactor $\theta_{13}$ experiments are under construction. Details of these experiments are discussed in this conference.[32] The expected sensitivities of these experiments range between 0.01 and 0.03 in $\sin^2 2\theta_{13}$.

The accelerator long baseline experiments try to observe $\nu_e$ appearance. See Eq. 3 for the approximate appearance probability. The T2K experiment[33] and the NOvA experiment[34] have high sensitivities to the $\nu_e$ appearance.

T2K uses a high intensity, low energy ($E_\nu < 1$ GeV) neutrino beam produced by the 40 GeV PS in the J-PARC accelerator complex, which is under construction in Tokai, Japan. The designed beam power is 0.75 MW. The proton beam was successfully accelerated to 30 GeV as of this writing. The far detector is Super-Kamiokande. The baseline length is 295 km. The neutrino energy is tuned to the maximum oscillation energy. For $\Delta m_{23}^2 = 2.5 \times 10^{-3}$ eV$^2$, the maximum oscillation energy is 600 MeV. To produce high-intensity, narrow-band beam, the off-axis beam technique is used. The axis of the primary beam is displaced by 2.5 degrees from the direction to the neutrino detector. The T2K experiment will start taking data in 2009.

NOvA is the proposed long baseline experiment in the United States. This experiment plans to use the existing NuMI beam line at Fermilab. A beam power upgrade plan to about 1 MW is under investigation. The far detector is a 15 kton fully-active, finely-segmented liquid scintillator detector. The baseline length is 810 km. In order to get the high flux at the oscillation maximum, this experiment also uses the off-axis technique. The peak neutrino energy is about 2 GeV. The NOvA experiment is expected to start taking data in the early 2010's.

These experiments have similar sensitivities in $\sin^2 2\theta_{13}$. Assuming 5 years of data taking with the planned beam intensities, these experiments can find clear evidence for a non-zero $\theta_{13}$, if the true value of $\sin^2 2\theta_{13}$ is larger than $\sim 0.02$. If there is no evidence for a non-zero $\theta_{13}$, they can set the upper limit of $\sim 0.01$ on $\sin^2 2\theta_{13}$. The sensitivities of these long baseline experiments on $\sin^2 2\theta_{13}$ are similar to or slightly better than the next generation reactor $\theta_{13}$ experiments.

Finally it should be mentioned that these long baseline experiments can measure $|\Delta m_{23}^2|$ and $\sin^2 2\theta_{23}$ precisely. These experiments have the approximate sensitivities of 0.01 in $\sin^2 2\theta_{23}$. It is especially interesting if these experiments could find evidence for a non-maximal $\sin^2 2\theta_{23}$.

## 5.2. *Oscillation physics beyond $\theta_{13}$*

We assume that the next generation long baseline and reactor experiments will observe evidence for a non-zero $\theta_{13}$, suggesting $\sin^2 2\theta_{13}$ to be larger than $\sim 0.01$. In this case, we know the approximate values of the three mixing angles. However, we believe that this should not be the end of the neutrino oscillation studies. In fact, there are still important and unknown neutrino oscillation parameters. One is the CP phase. According to the leptogenesis scenario,[5] the seed of the baryon asymmetry in the Universe is generated from the CP violating decay of the heavy Majorana particles of the see-saw mechanism. Therefore, it is generally believed that the measurement of the CP violation in the neutrino sector is a very important step forward for the understanding of the baryon asymmetry in the Universe. In addition to the CP violation, we should note that we do not know whether the $\nu_3$ mass eigenstate is the heaviest or the lightest. This is also an important question to be addressed experimentally.

If $\theta_{13}$ is non-zero, future large-scale long-baseline experiments can address these questions. Various possibilities for these experiments have been studied. One possibility for such experiments could be the Phase-2 of the T2K project.[33] The CP violation phase can be measured by observing the difference in the neutrino oscillation probabilities between $\nu_\mu \to \nu_e$ and $\bar{\nu}_\mu \to \bar{\nu}_e$. To observe this effect, a huge detector (0.54 Mton fiducial mass, Hyper-Kamiokande), and a beam power upgrade of J-PARC will be required. Due to the relatively short baseline length of 295 km and therefore due to a relatively small matter effect, the mass hierarchy may not be determined in this setup. The T2K experiment uses an off-axis beam. It is noticed that, with the present T2K beam-line configuration, the beam is simultaneously available in Kamioka (which is 295 km away from the target) and Korea (which is more than 1000 km away from the target). Studies have been carried out assuming the two detectors in Kamioka and Korea with 0.27 Mton fiducial mass for both detectors.[35,36]

The other study has been carried out in the United States, which assumed a 1300 km baseline between Fermilab and DUSEL at Homestake.[37] The beam power could be as high as 2 MW. The detectors used in this study are a 300 kton water Cherenkov detector and a 100 kton liquid Argon detector.

From these studies, it is understood that the CP phase and mass hierarchy can be determined by these future experiments if $\sin^2 2\theta_{13}$ is larger than $\sim 0.01$ and if the baseline length is 1000 km or longer.

## 6. Summary

"Atmospheric neutrino anomaly" and "solar neutrino problem" were beautifully resolved by neutrino oscillations in 1998 and 2001(2), respectively. Since then, many experiments have contributed to the understanding of oscillations. Dominant $\nu_\mu \rightarrow \nu_\tau$ and $\nu_e \rightarrow (\nu_\mu + \nu_\tau)$ are already studied accurately. However, it is known that there are many interesting physics in the sub-dominant oscillations. They include $\theta_{13}$, the CP violated and the mass hierarchy. Therefore there are large activities to study these effects. The immediate goal is the search for non-zero $\theta_{13}$. Experiments to search for non-zero $\theta_{13}$ will start within a year or two. The subsequent big goals are the measurement of the CP violation and the determination of the mass hierarchy. The world community is working hard for the best strategy for these measurements.

## Acknowledgments

The author would like to thank the organizers of this conference for the kind invitation. This work was partly supported by the Grant-in-Aid in Scientific Research of JSPS.

## References

1. W. Pauli Septieme Conseil de Physique Solvay 1933: *Noyaux Atomiques*, Paris 1934, p. 324f.
2. P. Minkowski, *Phys. Lett. B* **67**, p. 421 (1977).
3. T. Yanagida in Proceedings of the Workshop on Unified Theories and Baryon

Number in the Universe, eds. O. Sawada and A. Sugamoto (KEK report 79-18, 1979).

4.  M. Gell-Mann, P. Ramond and R. Slansky in Supergravity, eds. P. van Nieuwenhuizen and D. Z. Freedman (North Holland, Amsterdam, 1979).
5.  M. Fukugita and T. Yanagida, *Phys. Lett. B* **174**, p. 45 (1986).
6.  Z. Maki, M. Nakagawa and S. Sakata, *Prog. Theor. Phys.* **28**, p. 870 (1962).
7.  B. Pontecorvo, *Sov. Phys. JETP* **26**, 984 (1968).
8.  K. S. Hirata *et al.*, *Phys. Lett. B* **205**, p. 416 (1988).
9.  K. S. Hirata *et al.*, *Phys. Lett. B* **280**, 146 (1992).
10. R. Becker-Szendy *et al.*, *Phys. Rev. D* **46**, 3720 (1992).
11. Y. Fukuda *et al.*, *Phys. Lett. B* **335**, 237 (1994).
12. Y. Fukuda *et al.*, *Phys. Rev. Lett.* **81**, 1562 (1998).
13. T. Kajita (For the Kamiokande and Super-Kamiokande collaborations), *Nucl. Phys. Proc. Suppl.* **77**, 123 (1999).
14. S. Fukuda *et al.*, *Phys. Rev. Lett.* **85**, 3999 (2000).
15. Y. Ashie *et al.*, *Phys. Rev. Lett.* **93**, p. 101801 (2004).
16. M. H. Ahn *et al.*, *Phys. Rev. D* **74**, p. 072003 (2006).
17. P. Adamson *et al.*, *Phys. Rev. Lett.* **101**, p. 131802 (2008).
18. R. Davis Jr., D. S. Harmer and K. C. Hoffman, *Phys. Rev. Lett.* **20**, 1205 (1968).
19. J. N. Bahcall and M. H. Pinsonneault, *Phys. Rev. Lett.* **92**, p. 121301 (2004).
20. L. Wolfenstein, *Phys. Rev. D* **17**, 2369 (1978).
21. S. P. Mikheev and A. Y. Smirnov, *Sov. J. Nucl. Phys.* **42**, 913 (1985).
22. K. S. Hirata *et al.*, *Phys. Rev. Lett.* **63**, p. 16 (1989).
23. P. Anselmann *et al.*, *Phys. Lett. B* **285**, 376 (1992).
24. D. N. Abdurashitov *et al.*, *Phys. Lett. B* **328**, 234 (1994).
25. Q. R. Ahmad *et al.*, *Phys. Rev. Lett.* **87**, p. 071301 (2001).
26. S. Fukuda *et al.*, *Phys. Rev. Lett.* **86**, 5651 (2001).
27. Q. R. Ahmad *et al.*, *Phys. Rev. Lett.* **89**, p. 011301 (2002).
28. K. Eguchi *et al.*, *Phys. Rev. Lett.* **90**, p. 021802 (2003).
29. T. Araki *et al.*, *Phys. Rev. Lett.* **94**, p. 081801 (2005).
30. T. Schwetz, M. Tortola and J. W. F. Valle, *New J. Phys.* **10**, p. 113011 (2008).
31. M. Apollonio *et al.*, *Eur. Phys. J. C* **27**, 331 (2003).
32. Y. Wang in these proceedings.
33. Y. Itow *et al.* (2001), hep-ex/0106019.
34. D. S. Ayres *et al.* (2004), hep-ex/0503053.
35. M. Ishitsuka, T. Kajita, H. Minakata and H. Nunokawa, *Phys. Rev. D* **72**, p. 033003 (2005).
36. F. Dufour (2008), Talk presented at the 4th International Workshop on Nuclear and Particle Physics at J-PARC (NP08), Mito, Japan.
37. V. Barger *et al.* (2007), arXiv:0705.4396 [hep-ph].
38. S. Abe *et al.* (KamLAND Collaboration), *Phys. Rev. Lett.* **100**, 221803 (2008).

# Chapter 2

# Future Atmospheric Neutrino Experiments: The Case of Water Cherenkov Detectors*

T. Kajita, M. Ishitsuka and K. Okumura

*Research Center for Cosmic Neutrinos, Institute for Cosmic Ray Research, Univ. of Tokyo, Kashiwa-no-ha 5-1-5, Kashiwa, Chiba 277-8582, Japan*
[†] *Presented at the workshop, kajita@icrr.u-tokyo.ac.jp*


Y. Obayashi and M. Shiozawa

*Kamioka Observatory, Institute for Cosmic Ray Research, Univ. of Tokyo, Mozumi, Kamioka, Gifu 506-1205, Japan*

ENSITIVITIES of future atmospheric neutrino experiments on neutrino oscillations are studied. Water Cherenkov detectors similar to Super-Kamiokande are assumed in this study. Emphases are made on the measurements of $\theta_{13}$ and the sign of $\Delta m^2_{23}$.

## 1. Introduction

Neutrino oscillations have been studied extensively, since studies of neutrino masses and mixing angles are on the few ways to explore physics beyond the standard model of particle physics.

[†]This article was originally published in *Neutrino Oscillations and Their Origin*, pp. 73–80 (2005).

Atmospheric neutrino experiments have been contributing to the study of neutrino oscillations. Early atmospheric neutrino experiments reported a smaller $\nu_\mu/\nu_e$ flux ratio than expected in the 1 GeV energy region.[1-5] In addition, zenith-angle dependent deficit of $\nu_\mu$ events was observed[6] for neutrinos in the multi-GeV energy range. Following these early studies, neutrino oscillation was discovered by Super-Kamiokande.[7] Consistent results have been obtained from other recent atmospheric neutrino experiments.[8,9] More detailed studies of neutrino oscillations have been continuing, improving our knowledge on the oscillation parameters[10] and excluding pure $\nu_\mu \leftrightarrow \nu_{sterile}$ oscillations[11,12] and non-standard oscillation models.[13,14] Preliminary results on $\nu_\tau$ appearance have also been reported.[15] In this workshop, the first evidence that the neutrino survival probability obeys the sinusoidal function as predicted by neutrino oscillations was presented based on an $L/E$ analysis.[16] The observed $L/E$ distribution disfavored the neutrino decay and decoherence models that could explain the zenith angle and energy dependent deficit of atmospheric $\nu_\mu$ events.

We note that a large fraction of these studies have been carried out by large water Cherenkov experiments. These results suggest that atmospheric neutrino experiments using the water Cherenkov technique should be able to contribute to the neutrino oscillation studies in the future as well. In this paper, we discuss neutrino oscillation physics with future atmospheric neutrino experiments based on the water Cherenkov technique. The following sections discuss parameter determination for pure $\nu_\mu \leftrightarrow \nu_\tau$ oscillations, search for non-zero $\theta_{13}$ and the determination of the sign of $\Delta m_{23}^2$.

## 2. Measurement of $\sin^2 2\theta_{23}$ and $\Delta m_{23}^2$

The mixing parameters, $\theta_{ij}$, are one of a few fundamental parameters that could constrain theories beyond the standard model of particle physics. Therefore, it is important to measure $\sin^2 2\theta_{23}$ as accurate as possible. The atmospheric neutrino flux is accurately predicted to be up-down symmetric in the multi-GeV energy range. The number of downward-going and upward-going $\nu_\mu$ events and $\sin^2 2\theta_{23}$ are related by $(up/down)_{Data}/(up/down)_{MC} \simeq 1 - (\sin^2 2\theta_{23})/2$ to first approximation, assuming that there is no (full) oscillation effect for downward-(upward-)going neutrinos. The systematic error in the up-down ratio measurement is about 0.9% (0.4% from the flux calculation, 0.7% from the up-down asymmetry in the energy calibration, and less than 0.4% from the possible background contamination) in the present Super-Kamiokande experiment.[19] Hence atmospheric neutrino data are useful for an accurate measurement of $\sin^2 2\theta_{23}$. The accuracy of the $\sin^2 2\theta_{23}$ measurement will be improved with (exposure time)$^{1/2}$ and will be about $\pm 2\%$ at $1\sigma$ for 1 Mton·yr exposure of a water Cherenkov detector.[17]

Super-Kamiokande has shown that $\Delta m_{23}^2$ can be measured accurately by measuring a dip in the $L/E$ distribution.[16] Figure 1 shows the expected sensitivity in $\sin^2 2\theta_{23}$ and $\Delta m_{23}^2$ by future atmospheric neutrino data with the $L/E$ analysis. Since the actual sensitivity depends strongly on the actual experimental conditions such as the detector size and the method of measuring the muon momentum, it is difficult to predict the general sensitivity as a function of the detector exposure. In addition, in the case of Super-Kamiokande, on which the figure is based, the sensitivity strongly depends on the true $\Delta m_{23}^2$ value, because the dip moves to a lower $L/E$ position (i.e., a higher neutrino energy for the same $L$) for a higher $\Delta m_{23}^2$. However, the number of events decreases for higher energy neutrinos. In addition, the acceptance for fully contained events decreases for higher energy neutrinos. Hence, the sensitivity in $\Delta m_{23}^2$ gets worse rapidly with higher $\Delta m_{23}^2$. We note that large atmospheric neutrino experiments could compete in the $\Delta m_{23}^2$ measurement with future long baseline neutrino experiments if the true $\Delta m_{23}^2$ value is near the lower edge of the presently allowed $\Delta m^2$ region, where the planned long baseline experiments have somewhat limited sensitivities.

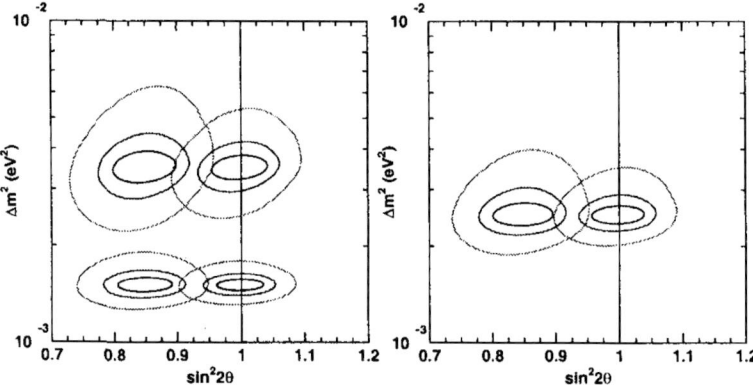

Figure 1.   Expected 90% C.L. allowed parameter regions by future atmospheric neutrino data on $\sin^2 2\theta_{23}$ and $\Delta m_{23}^2$ from an $L/E$ analysis with 110, 450 and 1800 kton·yr exposure of the detector. The Super-Kamiokande detector is assumed. Three different $\Delta m_{23}^2$ values (1.5, 2.5 and $3.5 \times 10^{-3}$ eV$^2$) and two $\sin^2 2\theta_{23}$ values (0.85 and 1.0) are assumed.

## 3. $\theta_{13}$

$\theta_{13}$ is a key parameter for the understanding of the neutrino mixing matrix. Therefore, various reactor and long-baseline accelerator experiments are designed to measure $\theta_{13}$. Atmospheric neutrino experiments have sensitivities in $\theta_{13}$ as well. Assuming that the effect of $\Delta m_{12}^2$ and $\theta_{12}$ can be neglected (a reasonable assumption for multi-GeV atmospheric neutrinos), the neutrino oscillation probability is written as;

$$P(\nu_e \rightarrow \nu_e) = 1 - s^2 2\theta_{13} s^2 \left( \frac{1.27 \Delta m_{23}^2 (\text{eV}^2) L (\text{km})}{E_\nu (\text{GeV})} \right), \qquad (1)$$

$$P(\nu_\mu \rightarrow \nu_e) = s^2 \theta_{23} s^2 2\theta_{13} s^2 \left( \frac{1.27 \Delta m_{23}^2 L}{E_\nu} \right), \qquad (2)$$

$$P(\nu_\mu \rightarrow \nu_\mu) = 1 - (s^4 \theta_{23} s^2 2\theta_{13} + c^2 \theta_{13} s^2 2\theta_{23}) s^2 \left( \frac{1.27 \Delta m_{23}^2 L}{E_\nu} \right), \quad (3)$$

where $s$ and $c$ represent sine and cosine, respectively. Since $\nu_e$ is involved in the oscillation, the matter effect[21,22] must be taken into account. The effect of a non-zero $\theta_{13}$ could be observed as an

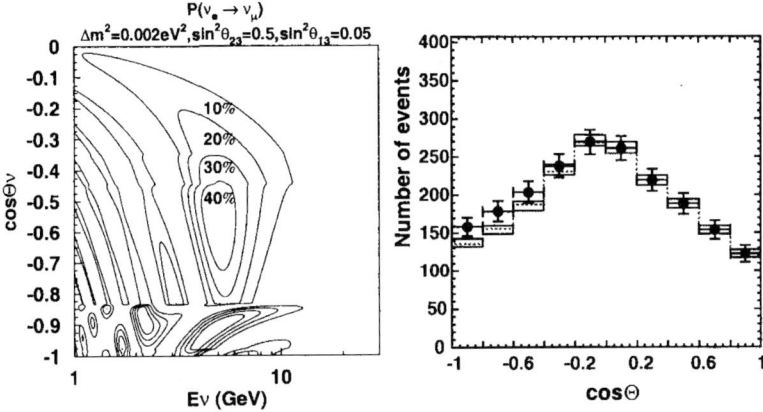

Figure 2. Left: $\nu_e \leftrightarrow \nu_\mu$ oscillation probability for neutrinos passing through the earth as a function of the neutrino energy and zenith angle for $\Delta m_{23}^2 = 2.0 \times 10^{-3}$ eV$^2$, $\sin^2 \theta_{23} = 0.50$ and $\sin^2 \theta_{13} = 0.05$. Right: Expected zenith angle distribution for single + multi-ring $e$-like events with the observed energy between 2.5 and 5 GeV for no oscillation (box histogram), and oscillation with $\Delta m_{13}^2 = 2.0 \times 10^{-3}$ eV$^2$, $\sin^2 \theta_{23} = 0.50$ and $\sin^2 \theta_{13} = 0.0$ (dotted histogram) and 0.05 (points). The height of the boxes and the vertical error bars show the statistical error in the MC simulation. The assumed detector exposure is 450 kton·yr.

excess of electron neutrinos in the upward-going direction through the matter resonance effect in the high energy range. For $\Delta m^2 = 2$ to $3 \times 10^{-3}$ eV$^2$, the resonance could occur for neutrinos with their energies between 5 and 10 GeV. Figure 2 (left) shows the $\nu_e \leftrightarrow \nu_\mu$ oscillation probability as a function of the neutrino energy and zenith angle. A clear resonance effect is seen for upward-going neutrinos near 5 GeV. For neutrinos passing through the core of the earth, resonances will occur in slightly lower neutrino energies.

Figure 2 (right) shows the expected excess of $e$-like events as a function of the zenith angle for $e$-like events with the energies between 2.5 and 5 GeV in a water Cherenkov detector. The assumed detector exposure is 450 kton·yr. In this figure, $\Delta m^2 = 2.0 \times 10^{-3}$ eV$^2$, and $\sin^2 \theta_{23} = 0.5$ and $\sin^2 \theta_{13} = 0.05$ (approximate CHOOZ limit[20]) are assumed. A statistically significant excess electron signal could be observed for these oscillation parameters.

Figure 3 shows the expected allowed regions in the $\sin^2 \theta_{13}$ vs. $\sin^2 \theta_{23}$ plane in cases where the true neutrino oscillation parameters

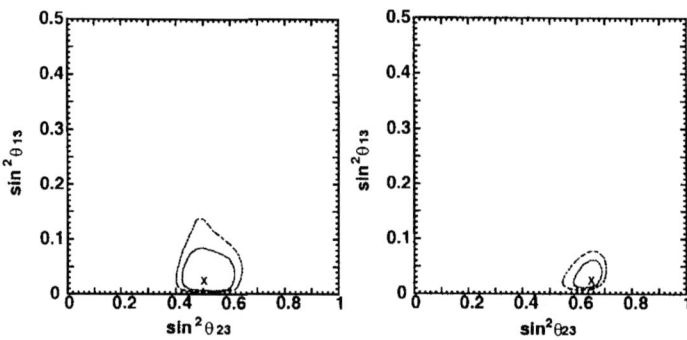

Figure 3. Expected 90 and 99% C.L. allowed parameter regions in the $\sin^2\theta_{13}$ vs. $\sin^2\theta_{23}$ plane after 450 kton·yr of the detector exposure for $\Delta m_{23}^2 = +2.5 \times 10^{-3}$ eV$^2$ (positive $\Delta m_{23}^2$), $\sin^2\theta_{13} = 0.025$ and $\sin^2\theta_{23} = 0.5$ (left) or 0.65 (right).

are $\Delta m_{23}^2 = 2.5 \times 10^{-3}$ eV$^2$, $\sin^2\theta_{13} = 0.025$ and $\sin^2\theta_{23} = 0.5$ (left) or 0.65 (right). $\sin^2\theta_{23} = 0.65$ (0.35) approximately corresponds to the present upper (lower) limit. The detector exposure is assumed to be 450 ton·yr. Fully contained, partially contained and upward-going muon events are used. The data are binned according to the observed energy (or momentum) and the zenith angle. Various systematic errors are included assuming that the systematic errors are identical to those of the Super-Kamiokande atmospheric neutrino analysis.[10,19] Several important features are visible in these figures. 90 and 99% C.L. allowed regions exclude $\theta_{13} = 0$ for the assumed parameter sets. 2 fold ambiguity in $\theta_{23}$ ($= 45° \pm \Delta\theta$) is resolved for sufficiently large $+\Delta\theta$. The allowed $\sin^2\theta_{13}$ region is smaller for larger $\theta_{23}$ values for a common $\theta_{13}$.

Figure 4 shows the expected $\chi^2$ difference between finite and null $\sin^2\theta_{13}$ assumptions for various $\sin^2\theta_{13}$, $\sin^2\theta_{23}$ and $\Delta m_{23}^2$ values. It is evident that the chance of observing finite $\theta_{13}$ increases for large $\sin^2\theta_{23}$.[18] It is also found that the sensitivity does not depend strongly on $\Delta m_{23}^2$. Because of the matter effect, the sensitivity slowly changes above $\sin^2\theta_{13} = 0.01$.

The resonance effect occurs only for neutrinos for positive $\Delta m^2$, and therefore only appears for the $e^-$ and $\mu^-$ spectrum. This, in

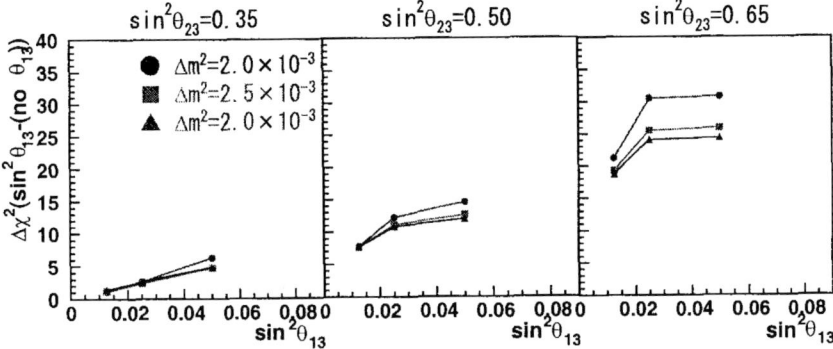

Figure 4. Expected $\chi^2$ difference between finite and null $\sin^2 \theta_{13}$ for $\Delta m_{23}^2 = +2.0$ (circle), 2.5 (square) and $3.0 \times 10^{-3}$ eV$^2$ (triangle) (positive $\Delta m_{23}^2$), and $\sin^2 \theta_{23} = 0.35$ (left), 0.5 (center) and 0.65 (right). The detector exposure is assumed to be 450 kton·yr.

turn, suggests that the sign of $\Delta m_{23}^2$ could be measured by atmospheric neutrino experiments. It is generally believed that a large magnetized atmospheric neutrino experiment is necessary to measure the sign of $\Delta m_{23}^2$,[23,24] while Super-Kamiokande and other water Cherenkov detectors are unable to distinguish $\nu_e$ and $\bar{\nu}_e$ interactions event-by-event bases. However, the cross section and the $y$ $(= (E_\nu - E_{lepton})/E_\nu)$ dependence of the cross section are different between $\nu$ and $\bar{\nu}$, and therefore it may be possible to distinguish the positive and negative $\Delta m_{23}^2$. For positive $\Delta m_{23}^2$, the resonance effect occurs only for neutrinos. Since the neutrino interactions produce more high-$y$ events (i.e., more multi-hadron events) than the antineutrino interactions, a larger effect of the finite $\theta_{13}$ can be seen in multi-ring $e$-like events for positive $\Delta m_{23}^2$ than for negative $\Delta m_{23}^2$. Figure 5 shows the sensitivity of a water Cherenkov atmospheric neutrino detector in the measurement of the sign of $\Delta m_{23}^2$. It is possible to measure the sign of $\Delta m_{23}^2$ in water Cherenkov detectors, if the $\sin^2 \theta_{13}$ and $\sin^2 \theta_{23}$ values are respectively near the present limit and $\geq 0.5$, provided that the detector exposure is more than 1 Mton·yr. Upper and lower figures in Fig. 5 assume that the true $\Delta m_{23}^2$ is positive and negative, respectively. In the case of negative $\Delta m_{23}^2$, the sensitivity is slightly worse.

*Experimental Studies of Neutrino Oscillations*

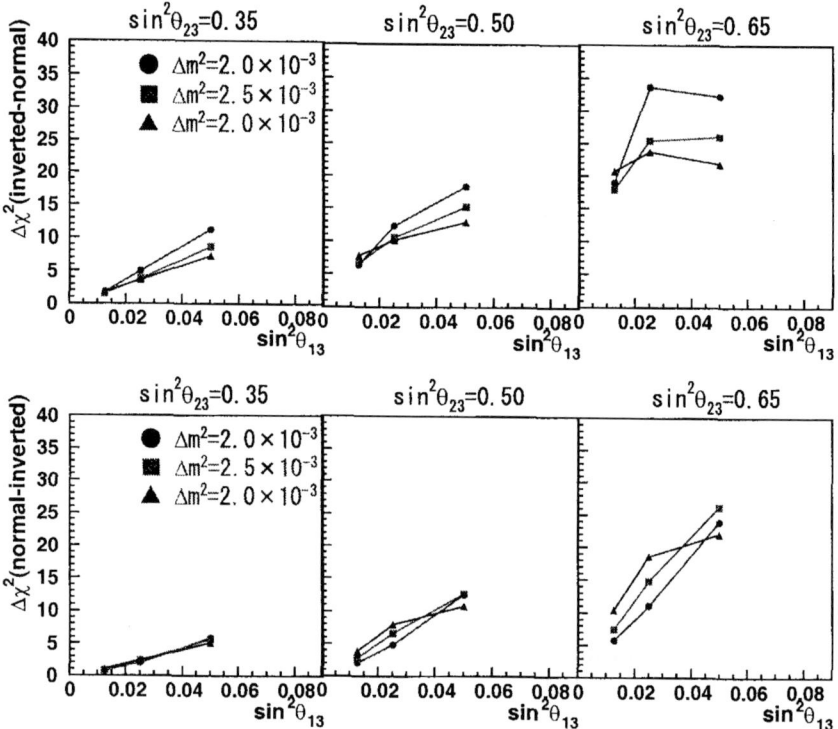

Figure 5. Expected $\chi^2$ difference for positive and negative $\Delta m^2$ assumptions for $\sin^2 \theta_{13} = 0.0125$, $0.025$ and $0.05$, $\Delta m^2_{23} = 2.0$ (circle), $2.5$ (square) and $3.0 \times 10^{-3}$ eV$^2$ (triangle), and $\sin^2 \theta_{23} = 0.35$ (left), $0.5$ (center) and $0.65$ (right). The detector exposure is assumed to be 180 Mton·yr. The true $\Delta m^2_{23}$ is assumed to be positive and negative for the upper and lower figures, respectively.

## 4. Effects of the Solar Oscillation Terms

The present study does not include the oscillation terms that are related to solar neutrinos ($\theta_{12}$ and $\Delta m^2_{12}$). It has been pointed out that these terms could play unique roles to the atmospheric neutrino oscillations, such as the possible measurement of $\sin^2 \theta_{23}$, (i.e., the discrimination of $\theta_{23} > 45°$ and $< 45°$).[25] These effects should be studied seriously taking various systematic errors into account.

## 5. Summary

We have studied the sensitivities of future water Cherenkov atmospheric neutrino experiments. The accuracy of the $\Delta m_{23}^2$ and $\sin^2 2\theta_{23}$ measurement is likely to improve steadily with the detector exposure. The future $L/E$ analysis could be very important for the determination of $\Delta m_{23}^2$, if the true $\Delta m_{23}^2$ is near the present lower limit, where the planned long baseline experiments have limited sensitivities. Furthermore, atmospheric neutrino experiments could contribute to the measurements of $\theta_{13}$ and the sign of $\Delta m_{23}^2$, provided that the true $\sin^2 \theta_{13}$ is near the present limit. It is likely that atmospheric neutrino experiments will continue to contribute to the neutrino oscillation studies.

## Acknowledgments

We would like to thank the members of the Super-Kamiokande collaboration for useful discussions. Also, we would like to thank A. Smirnov, S. Petcov and P. Lipari for stimulating discussions at the NOON2004 workshop. This work has been supported by the Japanese Ministry of Education, Culture, Sports, Science and Technology.

## References

1. K. S. Hirata *et al.*, *Phys. Lett. B* **205**, 416 (1988).
2. K. S. Hirata *et al.*, *Phys. Lett. B* **280**, 146 (1992).
3. D. Casper *et al.*, *Phys. Rev. Lett.* **66**, 2561 (1991).
4. R. Becker-Szendy *et al.*, *Phys. Rev. D* **46**, 3720 (1992).
5. W. W. M. Allison *et al.*, *Phys. Lett. B* **391**, 491 (1997).
6. Y. Fukuda *et al.*, *Phys. Lett. B* **335**, 237 (1994).
7. Y. Fukuda *et al.*, *Phys. Rev. Lett.* **81**, 1562 (1998).
8. M. Sanchez *et al.*, *Phys. Rev. D* **68**, 113004 (2003).
9. M. Ambrosio *et al.*, *Phys. Lett. B* **566**, 35 (2003).
10. C. Saji, for the Super-Kamiokande collaboration, *Neutrino Oscillations and Their Origin* (2005).
11. S. Fukuda *et al.*, *Phys. Rev. Lett.* **85**, 3999 (2000).

12. M. Ambrosio *et al.*, *Phys. Lett. B* **517**, 59 (2001).
13. P. Lipari and M. Lusignoli, *Phys. Rev. D* **60**, 013003 (1999).
14. G. L. Fogli *et al.*, *Phys. Rev. D* **60**, 053006 (1999).
15. C. Saji, for the Super-Kamiokande collaboration, Proc. of the 28th International Cosmic Ray Conference, Vol. 3, p. 1267 (2003).
16. M. Ishitsuka, for the Super-Kamiokande collaboration, *Neutrino Oscillations and Their Origin* (2005); Y. Ashie *et al.*, *Phys. Rev. Lett.* in press, hep-ex/0404034.
17. T. Kajita, for the Super-Kamiokande collaboration, *Nucl. Phys. B (Proc. Suppl.)* **100**, 139 (2001).
18. J. Bernabeu, S. Palomares-Ruiz and S. T. Petcov, *Nucl. Phys. B* **669**, 255 (2003).
19. Y. Ashie *et al.*, draft in preparation.
20. M. Apollonio *et al.*, *Phys. Lett. B* **466**, 415 (1999).
21. S. P. Mikheyev and A. Yu. Smirnov, *Sov. J. Nucl. Phys.* **42**, 1441 (1985); *Nuovo Cimento C* **9**, 17 (1986).
22. L. Wolfenstein, *Phys. Rev. D* **17**, 2369 (1978).
23. A. Geiser, *Nucl. Phys. B (Proc. Suppl.)* **91**, 147 (2001).
24. Tommaso Tabarelli de Fatis, Proc. of La Thuile 2001, Results and perspectives in particle physics, La Thuile, Italy (2001) p. 677, hep-ph/0106252.
25. O. L. G. Peres and A. Yu. Smirnov, *Phys. Lett. B* **456**, 204 (1999); O. L. G. Peres and A. Yu Smirnov, *Nucl. Phys. B* **680**, 479 (2004).

# Chapter 3

# The JHF-Kamioka Neutrino Project*

Takaaki Kajita

*Research Center for Cosmic Neutrinos, Institute for Cosmic Ray Research, University of Tokyo, Kashiwa-no-ha 5-1-5, Kashiwa, Chiba 939-2255, Japan*
*kajita@icrr.u-tokyo.ac.jp*

THE JHF-Kamioka neutrino project is a second generation long baseline neutrino oscillation experiment that probes physics beyond the Standard Model by high precision measurements of the neutrino masses and mixing. A high intensity neutrino beam is produced by a high intensity proton synchrotron at JHF. The neutrino energy is tuned to the oscillation maximum at $< 1$ GeV for the baseline length of 295 km. The far detector is the 50 kton water Cherenkov detector, Super-Kamiokande.

## 1. Introduction

The discovery of neutrino oscillations in atmospheric neutrinos[1] (see also Ref. 2 for earlier data) has opened a new window to study physics beyond the Standard Model through neutrino masses and mixing. The present data from Super-Kamiokande[3] show that the

---

*This article was originally published in *Neutrino Oscillations and Their Origin*, pp. 239–248 (2002).

mixing between the 2nd and the 3rd generation neutrinos ($\theta_{23}$) is consistent with the full mixing ($\sin^2 2\theta_{23} > 0.90$ at 90% C.L.). Also the difference of the neutrino mass squared is constrained well ($1.6 \times 10^{-3}$ eV$^2$ < $\Delta m_{23}^2$ < $3.6 \times 10^{-3}$ eV$^2$). In addition, the solar neutrino problem was proven to be due to neutrino oscillations by precise measurements by SNO[4] and Super-Kamiokande.[5] The mixing between the first and the second generation neutrinos ($\theta_{12}$) is also found to be large. These neutrino mixing angles that are significantly different from those of quarks may be a key to a deeper understanding of the physics beyond the Standard Model. However, we know that our understanding of the neutrino mixing matrix is not complete, because we have little knowledge on the first and the third generation mixing ($\theta_{13}$).

Results from the first generation long baseline neutrino oscillation experiment, K2K,[6] are also consistent with the atmospheric neutrino data. In a few years, the MINOS experiment[7] will start taking data. This experiment will significantly improve our knowledge on $\sin^2 2\theta_{23}$ and $\Delta m_{23}^2$. The CNGS project[8] will also start experiments in 2005. The main goal of this project is the confirmation of $\nu_\mu \rightarrow \nu_\tau$ neutrino oscillations by direct detection of $\tau$ leptons.

The main goals of the JHF-Kamioka neutrino project are observation of non-zero $\sin^2 2\theta_{13}$ and precise measurement of $\sin^2 2\theta_{23}$ and $\Delta m_{23}^2$.

## 2. Overview of the Experiment

The JHF-Kamioka neutrino project is a proposed long baseline neutrino oscillation experiment using the JHF 50 GeV proton synchrotron (PS). The JHF accelerator complex was officially approved in 2001 by the Japanese government. The construction of the 50 GeV PS is in progress in JEARI (Japan Atomic Energy Research Institute) at Tokai, and will be completed in 2006. In the earliest case, the neutrino project can start in 2007. 5 years of experimental period is assumed for the first phase of this neutrino project. The 50 GeV PS is designed to deliver $3.3 \times 10^{14}$ protons every 3.4 seconds. The beam power is 0.77 MW. A future upgrade of the beam power to 4 MW is considered. The far detector is the Super-Kamiokande 50 kton water Cherenkov detector. The baseline length of the experiment is 295 km. A 1 Mton water Cherenkov detector, Hyper-Kamiokande, is seriously considered as a far detector in the second phase of this neutrino project.

A feature of this experiment is the use of a low-energy, narrow band, high-intensity neutrino beam. The neutrino energy will be tuned to the maximum oscillation energy. For $\Delta m_{23}^2 = 3.0 \times 10^{-3}$ eV$^2$, it is 715 MeV. In the early stage of the studies of this project, three different neutrino beams were considered. They are wide band beam (WBB), narrow band beam (NBB) by selecting mono-energetic pions, and off-axis beam (OAB). OAB is another option to produce a narrow neutrino energy spectrum.[9] The beam optics is almost same as the WBB, but the axis of the beam is displaced by a few degrees from the far detector direction. Due to the two body decay kinematics of pions, the energy of neutrinos that hit the far detector is almost independent of the pion energy spectrum. The neutrino energy can be adjusted by choosing the angle between the pion beam direction and the direction to the detector (off-axis angle).

Detailed Monte Carlo simulations have been carried out to estimate the expected neutrino spectrum and the number of events. Figure 1 shows a comparison of the expected spectra by the three beam configurations. It is clear that the OAB has the highest flux at the maximum oscillation energy. Table 1 shows the expected number of events for no oscillations. From these studies, we have decided to the use the OAB for the JHF-Kamioka neutrino project.

If the off-axis angle is determined, the neutrino energy distribution is essentially determined. On the other hand, our knowledge on $\Delta m_{23}^2$ is not precise enough to pre-determine the off-axis angle uniquely. Because of these conditions, the decay pipe is designed to accommodate off-axis angles between 2 and 3 degrees.

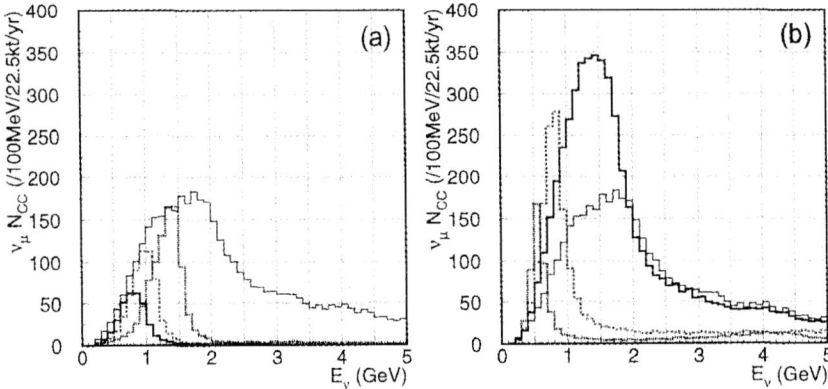

Figure 1. Neutrino energy spectra of charged current interactions. Thick solid, dashed and dotted histograms in (a) show NBB with 1.5, 2.0 and 3.0 GeV beam, and those in (b) show OAB with 1.0°, 2.0° and 3.0° off-axis angles, respectively. WBB is shown by a thin solid histogram in both (a) and (b).

Table 1. Summary of the expected number of events per year for various beam options. The oscillation effect is not included.

| Beam | $\nu_\mu$ CC | $\nu_e$ CC | NC | Total / year |
|---|---|---|---|---|
| Wide band beam | 5200 | 59 | 1800 | 7100 |
| Off-axis 2 degree | 2200 | 45 | 900 | 3200 |
| Off-axis 3 degree | 800 | 22 | 300 | 1100 |
| Narrow band beam ($E_\pi = 2$ GeV) | 620 | 5.0 | 250 | 880 |

## 3. The Neutrino Flux and the Near Detector

Another important element of the experiment is the front detector that monitors the neutrino flux without the oscillation effect. In order to achieve the designed goals of the experiment, it is very important to precisely understand the neutrino beam and the detection efficiency for various types of neutrino events. For this purpose, it is important that the near and the far detectors are designed as similar as possible so that various detector systematics cancel by taking the far-near ratio. The far detector, Super-Kamiokande, is a water Cherenkov detector. Therefore, we strongly think that the near detector should also be a water Cherenkov detector. By an event rate consideration, we estimate that the fiducial mass of the near detector should be about 100 tons, assuming that the detector is installed at 2 km from the target. We also require that the distance from the surface of the fiducial volume to the photomultiplier tubes (PMTs) is 2.0 m following the definition in Super-Kamiokande. In addition, the near detector has to measure the energy of muons up to about 1 GeV. For this reason, the distance from the surface of the fiducial volume to the PMTs should be about 5 m for the down-stream region. The fiducial volume can be 4 m in diameter and 8 m in length, and the total volume (including the volume for installing PMTs) can be about 9 m in diameter and 16 m in length. The total weight is about 1000 tons. In addition to the water Cherenkov detector, it might be important to install muon range counters at the down stream to measure the energies of high energy muons that shall be produced by the high energy tail of the flux. Furthermore, a fine-grained scintillator detector should be installed in front of the water Cherenkov detector in order to study the details of neutrino interaction kinematics.

It is known that the energy spectrum of the neutrinos at the Super-Kamiokande detector, which is 295 km away from the neutrino production point, shall not be identical to that at the front detector. For example, Figure 2 compares the calculated energy spectrum at Super-Kamiokande and at 0.28 and 1.5 km distances from the neutrino production target for the 2 degree off-axis beam. 0.28 km is the possible on-site neutrino detector position. The spectrum has lower energy peak at 0.28 km, while the 1.5 km and Super-Kamiokande spectra are almost identical. According to the beam Monte Carlo, the "far-near" ratio ($\equiv$ flux(far)/flux(near) $\times$ $(L_{far}/L_{near})^2$) approaches to unity very quickly as $L_{near}$ increases, where $L_{far}$ ($L_{near}$) is the distance between the production target and the far (near) detector. For $L_{near} > 1.0$ km, the deviation from unity is less than 10%.

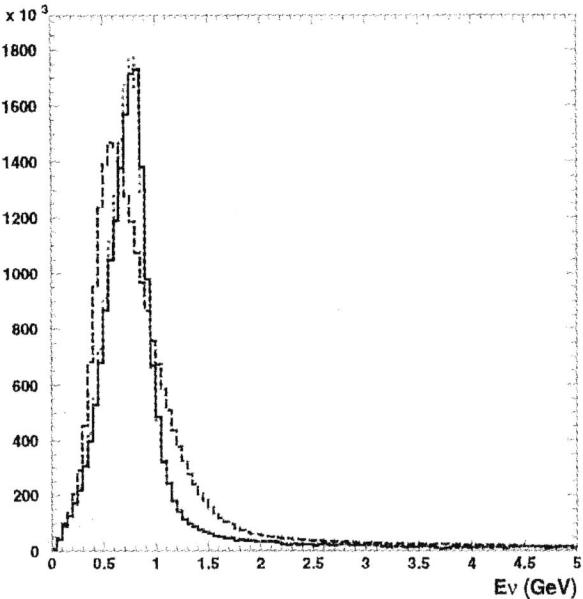

Figure 2. Calculated energy spectrum of the neutrino flux at Super-Kamiokande (solid) and at 0.28 (dashed) and 1.5 km (dotted) distances from the neutrino production target for the 2 degree off-axis beam. The vertical axis is arbitrary.

Even if the uncertainty in the far-near ratio is large, it might be acceptable if it does not affect the physics results. From various considerations, 5% far-near systematic error could be a goal for various measurements. Let us discuss the physics sensitivities. In the following discussions, we assume that the true $\Delta m_{23}^2$ is $3.0 \times 10^{-3}$ eV$^2$. We also assume the 2 degree off-axis beam.

## 4. Physics

### 4.1. *Measurement of* $\sin^2 2\theta_{23}$ *and* $\Delta m^2$

The neutrino energy can be reconstructed for quasi-elastic interactions assuming that the target nucleon is at rest;

$$E_\nu = \frac{m_N E_l - m_l^2/2}{m_N - E_l + P_l \cos\theta_l} \tag{1}$$

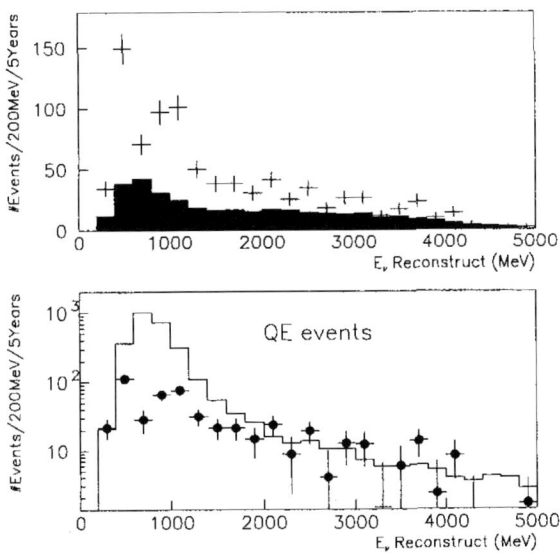

Figure 3. Upper: Reconstructed neutrino energy distribution for 5 year operation of the JHF-Kamioka experiment with the 2 degree off-axis beam. Events with single $\mu$-like Cherenkov ring are plotted. The shaded histogram shows the contribution of non-quasi-elastic events. Sin$^2 2\theta_{23} = 1.0$ and $\Delta m_{23}^2 = 3.0 \times 10^{-3}$ eV$^2$ are assumed. The maximum oscillation is expected to occur at the 0.6 to 0.8 GeV energy bin. Lower: Reconstructed $E_\nu$ distribution after subtracting the non-quasi-elastic events. The histogram and the dots show the non-oscillation and oscillation cases.

Figure 3 shows the expected energy spectrum with and without neutrino oscillations assuming sin$^2 2\theta_{23} = 1.0$ and $\Delta m_{23}^2 = 3.0 \times 10^{-3}$ eV$^2$. Events with single $\mu$-like Cherenkov ring are plotted. Since the peak flux is tuned to the maximum oscillation energy, most of the muon neutrinos must be oscillated to tau neutrinos. From this figure it is possible to estimate the sensitivities in sin$^2 2\theta_{23}$ and $\Delta m_{23}^2$. See Figure 4 (upper). The expected accuracy is 1% for sin$^2 2\theta_{23}$ and $1 \times 10^{-4}$ eV$^2$ for $\Delta m_{23}^2$, if the true sin$^2 2\theta_{23}$ value is 1.0. However, at the preparation stage we must also consider a case that the true sin$^2 2\theta_{23}$ is less than 1.0. (The present lower limit from Super-Kamiokande is 0.90 at 90% C.L.) The maximum oscillation effect should occur at the 600 to 800 MeV energy bin for $\Delta m_{23}^2 = 3 \times 10^{-3}$ eV$^2$. Since this bin is the most important one for

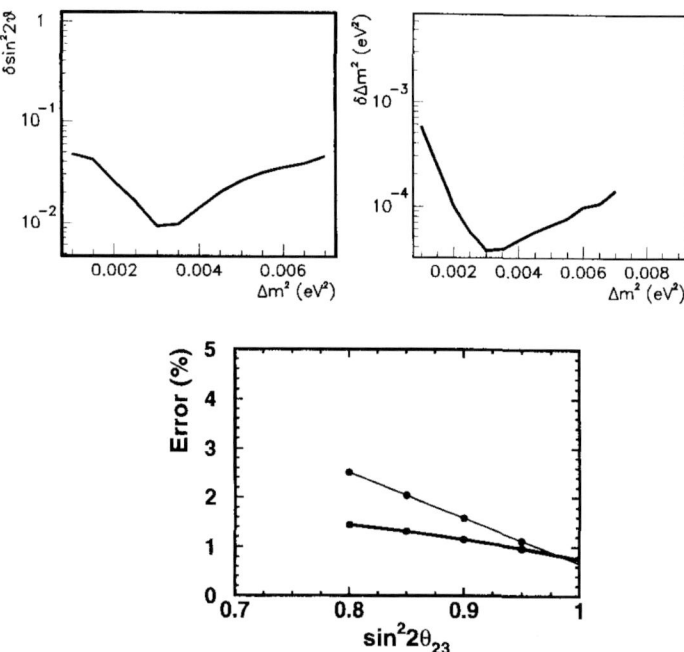

Figure 4. Upper: Sensitivity in $\sin^2 2\theta_{23}$ and $\Delta m^2_{23}$ after 5 years of operation of the JHF-Kamioka experiment with the 2 degree off-axis beam. Lower: Expected statistical (thick line) and systematic (thin line) errors in the $\sin^2 2\theta_{23}$ measurement after 5 years of operation of the experiment for various values of $\sin^2 2\theta_{23}$. 10% far-near ratio uncertainty is assumed for the systematic error.

the determination of $\sin^2 2\theta_{23}$, the statistical and systematic errors are estimated based on the number of events in this bin. Figure 4 (lower) shows the estimated statistical and systematic errors in the measurement of $\sin^2 2\theta_{23}$. If the far-near systematic error is assumed to be 10%, the statistical error is larger than the systematic error at $\sin^2 2\theta_{23} = 1.0$, and the total error is about 1%. However, for smaller $\sin^2 2\theta_{23}$ values, systematic error dominates the statistical error. Because the systematic error is proportional to the number of observed events in the 600 to 800 MeV energy bin, the systematic error rapidly gets bigger for smaller $\sin^2 2\theta_{23}$. In order not to limit the measurement by the far-near systematic error, it is essential to understand the far-near systematic effect to an accuracy of better than 5%.

## 4.2. *Search for non-zero* $\sin^2 2\theta_{13}$

The JHF neutrino beam has a small $\nu_e$ contamination (0.2% at the energy of the peak flux). Furthermore, the $\nu_e$ appearance signal is maximized by tuning the neutrino energy at its oscillation maximum. Thus, this experiment has an excellent opportunity to discover non-zero $\theta_{13}$.

The signal should be searched for in the single-ring *e*-like events. The possible background processes are single muon events with the muon being misidentified as an electron, contamination of $\nu_e$ in the beam, and NC (mostly $\pi^0$) events. Among them, the most serious one is the NC background. Special cuts have been developed to reject these events (see Ref. 10 for details of the cuts). Figure 5 shows the expected energy distribution for the signal and background. Table 2 summarizes the expected number of signal and background events for two different values of $\sin^2 2\theta_{13}$. The sensitivity of this experiment to $\sin^2 2\theta_{13}$ is about 0.006. In this paper, we also consider a case that the true $\sin^2 2\theta_{13}$ is relatively large and that the $\sin^2 2\theta_{13}$ value is measurable. From Table 2, it is easy to roughly estimate the statistical and systematic errors. Again, only the far-near ratio

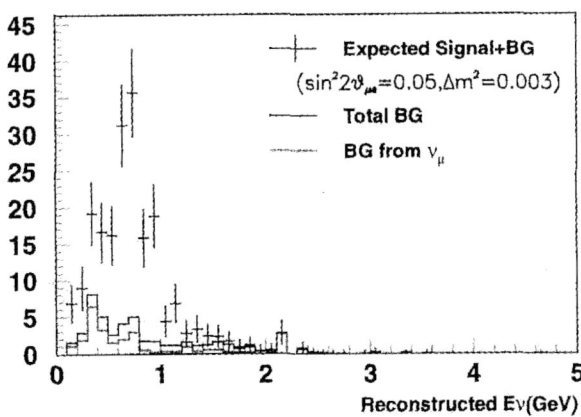

Figure 5. Expected electron appearance signal in the JHF-Kamioka neutrino project. It is assumed that $\sin^2 2\theta_{13} = 0.1$ and 5 years of operation. The matter and the CP violation effects are neglected.

Table 2. Summary of the expected electron appearance signal and background for the measurement of $\sin^2 2\theta_{13}$. $\Delta m_{13}^2 = 3.0 \times 10^{-3}$ eV$^2$ and $\sin^2 2\theta_{23} = 1.0$ are assumed. The matter and the CP violation effects are neglected.

| $\sin^2 2\theta_{13}$ | $\nu_\mu$(CC+NC) | Beam $\nu_e$ | Osc'd $\nu_e$ | Total (signal+BG) |
|---|---|---|---|---|
| 0.1 | 11.1 | 11.1 | 123.2 | 145.5 |
| 0.01 | 11.1 | 11.1 | 12.3 | 34.5 |

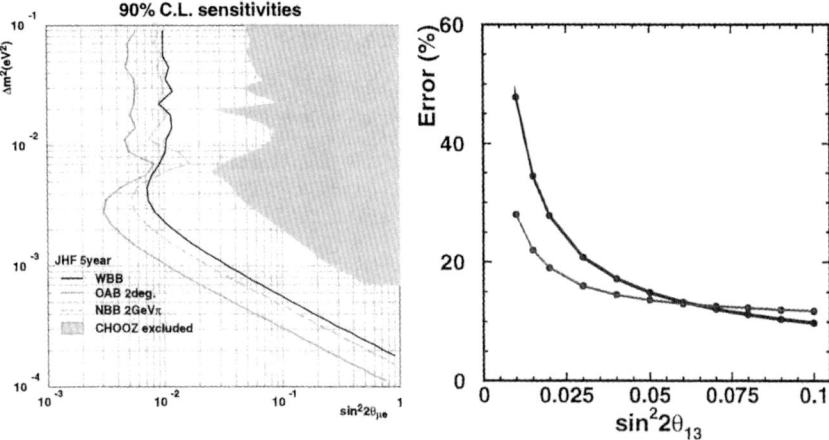

Figure 6. Left: 90% C.L. sensitivity in $\sin^2 \theta_{\mu e} \equiv 0.5 \cdot \sin^2 2\theta_{13}$ after 5 years of operation. Right: Expected statistical (thick line) and systematic (thin line) errors in the $\sin^2 2\theta_{13}$ measurement after 5 years of the operation of the JHF-Kamioka experiment for various values of $\sin^2 2\theta_{13}$. 10% far-near ratio uncertainty is assumed for the systematic error. Intrinsic uncertainties due to CP violation and sign of $\Delta m_{23}^2$ are neglected.

uncertainty is considered for the systematic error. Figure 6 (left) shows the expected sensitivity in $\sin^2 2\theta_{\mu e}$ ($\equiv 0.5 \times \sin^2 2\theta_{13}$). Figure 6 (right) shows the estimated statistical and systematic errors for various $\sin^2 2\theta_{13}$ values, where the far-near systematic error of 10% is assumed. $\sin^2 2\theta = 0.01$ and 0.1 approximately correspond to the sensitivity of this experiment and the lower limit from the CHOOZ[11] and Palo Verde[12] experiments, respectively. In order not to limit the measurement by the systematics, the far/near systematic error must be controlled to an accuracy of better than 5%.

## 5. Phase 2 of the Experiment

In the 2nd phase of the JHF-Kamioka neutrino project, the proton intensity delivered by the 50 GeV PS is planned to increase to 4 MW. Also a 1 Megaton water Cherenkov detector, Hyper-Kamiokande, is considered as the far detector. With these upgrades, it is possible to measure the CP phase in the lepton sector, provided that the solution of the solar neutrino problem is the Large Mixing Angle solution. However, I will not discuss in any details, because M. Aoki[13] discussed details of the CP violation physics with the JHF-Kamioka neutrino project.

## 6. Recent Progress

In this article, I have discussed the physics sensitivity based on Ref. 10. We feel that the event rate is not high enough for various measurements. Therefore, after writing Ref. 10, we have studied a possibility of a longer decay pipe for obtaining a higher flux. In Ref. 10, the length of the decay pipe was assumed to be 80 m. Based on the recent study, we decided to use 130 m decay pipe. The flux increased by about 40% at the peak energy of the flux. Therefore, the physics sensitivity described in this article could be achieved in about 3.5 years of operation (or a better sensitivity is expected for 5 years of operation).

## 7. Summary

The JHF-Kamioka neutrino project is a second generation neutrino oscillation experiment. The goals in the first phase are a precise measurement of $\sin^2 2\theta_{23}$ and $\Delta m_{23}^2$, and a discovery of non-zero $\theta_{13}$. The expected accuracy of the measurement is $\delta(\sin^2 2\theta_{23}) \sim 0.01$ and $\delta(\Delta m_{23}^2) \sim 1 \times 10^{-4}$ eV$^2$. If $\sin^2 2\theta_{13}$ is larger than 0.01 it is possible to observe non-zero $\theta_{13}$. In the second phase of this project, it is possible to measure the CP violation phase, provided that the solution to the solar neutrino problem is the Large Mixing Angle solution.

This project is a large project. It is essential that the experimental team is formed by physicists from all over the world who are interested in this project.

## Acknowledgments

The author would like to thank the members of the JHF-Kamioka neutrino working group for many useful discussions and suggestions.

## References

1. Y. Fukuda *et al.*, *Phys. Lett. B* **433**, 9 (1998); *Phys. Lett. B* **436**, 33 (1998); *Phys. Rev. Lett.* **81**, 1562 (1998); *Phys. Rev. Lett.* **82**, 2644 (1999); *Phys. Lett. B* **467**, 185 (1999); S. Fukuda *et al.*, *Phys. Rev. Lett.* **85**, 3999 (2000).
2. K. S. Hirata *et al.*, *Phys. Lett. B* **205**, 416 (1988); *Phys. Lett. B* **280**, 146 (1992); Y. Fukuda *et al.*, *Phys. Lett. B* **335**, 237 (1994); S. Hatakeyama *et al.*, *Phys. Rev. Lett.* **81**, 2016 (1998).
3. M. Messier, for the Super-Kamiokande collaboration, *Neutrino Oscillations and Their Origin* (2002).
4. Q. R. Ahmad *et al.*, *Phys. Rev. Lett.* **87**, 071301 (2001); S. Oser, for the SNO collaboration, *Neutrino Oscillations and Their Origin* (2002).
5. S. Fukuda *et al.*, *Phys. Rev. Lett.* **86**, 5651 (2001); *Phys. Rev. Lett.* **86**, 5656 (2001); Y. Koshio, for the Super-Kamiokande collaboration, *Neutrino Oscillations and Their Origin* (2002).
6. S. H. Ahn *et al.*, *Phys. Lett. B* **511**, 178 (2001); J. Hill, for the K2K collaboration, *Neutrino Oscillations and Their Origin* (2002).
7. M. Messier, for the MINOS collaboration, *Neutrino Oscillations and Their Origin* (2002).
8. M. Campanelli, *Neutrino Oscillations and Their Origin* (2002).
9. D. Beavis *et al.*, Proposal, BNL AGS E-889 (1995).
10. Y. Itow *et al.*, 2001, Letter of Intent "The JHF-Kamioka Neutrino Project", hep-ex/0106019.
11. M. Apollonio *et al.*, *Phys. Lett. B* **420**, 397 (1998); *Phys. Lett. B* **466**, 415 (1999).
12. F. Boehm *et al.*, *Phys. Rev. Lett.* **84**, 3764 (2000); *Phys. Rev D* **62**, 072002 (2000).
13. M. Aoki, *Neutrino Oscillations and Their Origin* (2002).

# Chapter 4

# Solar and Atmospheric Neutrino Results from Super-Kamiokande*

T. Kajita

For the Super-Kamiokande Collaboration

*Research Center for Cosmic Neutrinos, Institute for Cosmic Ray Research, Univ. of Tokyo, 5-1-5, Kashiwa-no-ha, Kashiwa, Chiba 277-8582, Japan kajita@icrr.u-tokyo.ac.jp*

RECENT data from Super-Kamiokande on solar and atmospheric neutrinos are presented. Their constraints on neutrino mass and mixing are discussed. The present solar neutrino data do not show any evidence for a day-night effect nor a energy spectrum distortion. These results disfavor the small-mixing MSW and vacuum oscillation solutions. The atmospheric neutrino data give accurate information on neutrino mass and mixing. The data are explained by two flavor $\nu_\mu \rightarrow \nu_\tau$ oscillations with $\Delta m^2 = (1.5 - 5) \times 10^{-3}$ eV$^2$ and $\sin^2 2\theta > 0.88$.

## 1. Introduction

Measurement of neutrino mass and mixing is one of a few ways to explore physics beyond the standard model of the particle physics. It has been known that observed solar and atmospheric neutrino

---

*This article was first published in *The Ninth Marcel Grossmann Meeting*, pp. 203–218 (2002).

data disagreed with theoretical predictions. Neutrino oscillations, and therefore neutrino mass and mixing, have been quoted as the explanations to these data. At present, it is recognized that the atmospheric neutrino data give evidence for $v_\mu \rightarrow v_\tau$ oscillations, and therefore the atmospheric neutrino field is maturing from the discovery phase to the detailed study phase. In this paper, recent atmospheric and solar neutrino data from Super-Kamiokande, which have been contributing to the neutrino oscillation studies, will be presented.

Super-Kamiokande is a 50 kton water Cherenkov detector which is located at a depth of 2700 meters water equivalent underground at the Kamioka Observatory, Kamioka, Japan. The detector is consisted of two parts, an inner detector and an outer detector. In the inner and outer detectors, Cherenkov photons are detected by 11146 and 1885 Photo Multiplier Tubes (PMTs), respectively. The charge and timing information recorded by the inner detector PMTs are used to reconstruct kinematics of neutrino events. The outer detector, which surround the inner detector completely by about 2 meters of water, is useful to identify incoming cosmic ray muons and exiting particles in neutrino events which occurred in the inner detector. The outer detector water passively shields against radioactivities (gammas and neutrons) from the surrounding rock. The fiducial volume for various physical analyses is 22.5 kton. The detector has been taking data since the spring of 1996. An event trigger is formed by a coincidence of the inner detector PMT hits with the hit multiplicity larger than a preset number. The trigger threshold at the beginning of the experiment was 5.7 MeV at a 50% efficiency. Since then, we have continuously tried to lower the threshold. The threshold is 3.5 MeV at a 50% efficiency as of this writing.

## 2. Solar Neutrinos

The basic mechanism of the energy generation in the Sun has been confirmed to be nuclear fusion processes by the detection of solar neutrinos by six solar neutrino experiments.[1-6] However, observed fluxes from these experiments were lower than the Standard Solar Model (SSM) prediction.[7] It is known that reasonable modifications of the solar model are essentially impossible to explain the present solar neutrino data. ON the other hand, it is possible to explain the solar neutrino data by neutrino oscillations in the Sun[8] or in the vacuum. For a definite confirmation of the solar neutrino oscillations, experimental results which can not be predicted by any solar models are highly desirable. Furthermore, the parameter region of neutrino oscillations should be uniquely determined. There are four different parameter regions (solutions) which can explain the existing solar neutrino data. It is predicted that a spectrum of the $^8$B solar neutrinos should be distorted significantly for the cases of the "small-mixing MSW solution" and the "vacuum oscillation solution", or a day-night effect should be observed for a region of the "large-mixing MSW solution".

Super-Kamiokande detects $^8$B solar neutrinos through $\nu e \rightarrow \nu e$. The electron energy, direction and time of the reaction are measured. The energy scale, energy resolution, angular resolution and vertex position resolution were calibrated mainly by an electron LINAC system which produces an electron beam of $5 \sim 16$ MeV energy.[9] In addition, $\beta$'s from $^{16}$N are used as an independent calibration source.[10] The absolute energy scale has been understood within 0.6% in this energy range. Recently the signal to noise ratio of the solar neutrino analysis has been improved by more than a factor of two relative to the previous analysis (see Figure 1). The measured $^8$B solar $\nu$ flux above 5.5 MeV during 1117 days of the observation time was $0.465 \pm 0.005$(stat.) $^{+0.015}_{-0.013}$(sys. of the data) of the SSM prediction.[7] The systematic error is about $\pm 3\%$. The main sources of the systematic error, which contribute to the total error by more than 1%, are uncertainties in the energy scale, the spectrum of $^8$B solar neutrinos, the reduction efficiency and the fiducial volume.

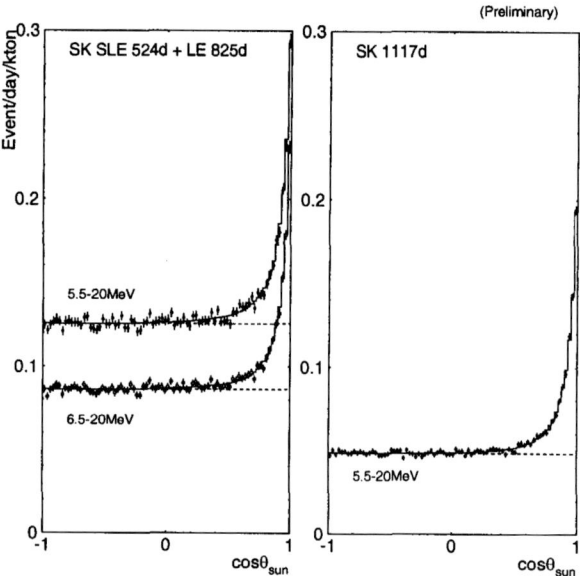

Figure 1. Solar neutrino signal observed in Super-Kamiokande. The left figure shows the previous data based on the older analysis. The right one shows the present data. The signal (a peak near $\cos\theta_{sun} = 1$) to noise (flat) ratio was improved by more than a factor of two.

Super-Kamiokande provides information for both the energy spectrum and the day-night effect. The day-night data are shown in Figure 2. The day-time and night-time fluxes were compared as: $(N - D)/((N + D)/2) = 0.034 \pm 0.022\text{(stat.)} {}^{+0.013}_{-0.012}\text{(sys.)}$. The possible excess of the flux in the night time was about 1.3 standard deviations and was not significant. Furthermore, we note that the data do not show any evidence for an enhancement of the day-night effect for neutrinos passing through the core of the Earth (see the N5 bin in Figure 2).

Figure 3 shows the energy spectrum of the recoil-electrons divided by the SSM prediction. In earlier Super-Kamiokande data, the Data/SSM values in the endpoint region were higher than the average. However, the excess is much less significant in the present data. A $\chi^2$ method was used to study a consistency of the energy spectrum with flat. The absolute flux value was assumed to be a

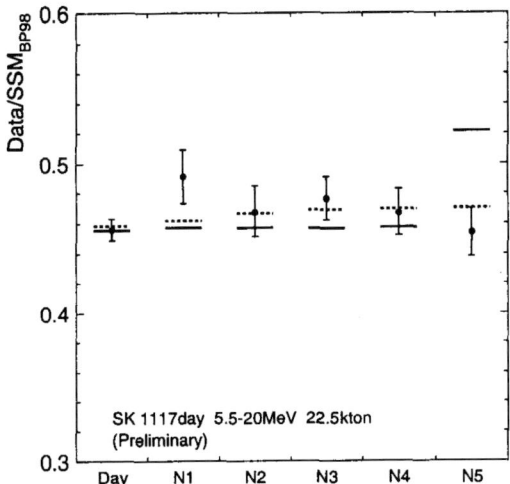

Figure 2.  Day-night data. The night time data are divided into 5 bins according to the relative direction to the Sun. Solid and dotted lines show the expected day-night effects for a small-mixing MSW solution ($\sin^2 2\theta = 0.008$, $\Delta m^2 = 8 \times 10^{-6}$ eV$^2$) and a large-mixing MSW solution ($\sin^2 2\theta = 0.7$, $\Delta m^2 = 6.3 \times 10^{-5}$ eV$^2$), respectively.

free parameter and only the shape of the distributions of the data and the expectations was compared. The $\chi^2$ values were obtained to be 13.7 for 17 degrees of freedom (DOF). The shape of the energy spectrum is consistent with the SSM prediction. Also shown in Figure 3 are expected shapes of the energy spectrum for several neutrino oscillation parameters. We find that the oscillation parameters which predict large energy-spectrum distortions are disfavored.

## 2.1. *Neutrino oscillation analysis*

The present day-night and energy spectrum data from Super-Kamiokande do not show any SSM independent evidence for neutrino oscillations. However, these data are useful to constrain neutrino oscillation parameters. In Figure 4 (left), black regions show 95% C.L. allowed neutrino oscillation parameter regions obtained by using the flux measurements from various solar neutrino experiments, assuming $\nu_e \rightarrow \nu_x$, where $\nu_x$ is either $\nu_\mu$ or $\nu_\tau$. Also

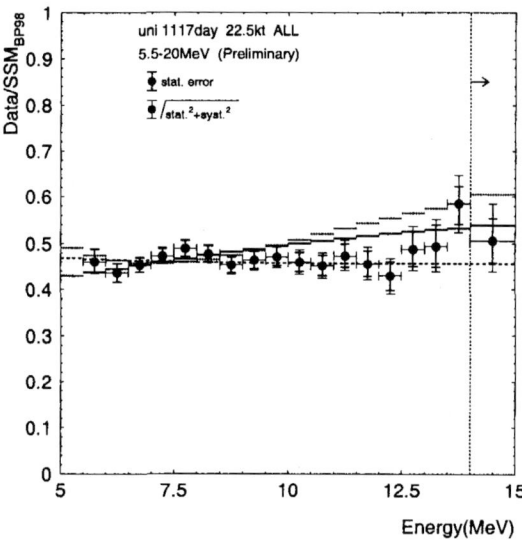

Figure 3.   Plot of Data/SSM as a function of the electron energy. Solid, dashed and dotted histograms show the expected energy spectrum for a typical small-mixing MSW solution ($\sin^2 2\theta = 6 \times 10^{-3}$, $\Delta m^2 = 5 \times 10^{-6}$ eV$^2$), a large-mixing MSW solution ($\sin^2 2\theta = 0.8$, $\Delta m^2 = 3.2 \times 10^{-5}$ eV$^2$), and a vacuum-oscillation solution ($\sin^2 2\theta = 0.75$, $\Delta m^2 = 6 \times 10^{-11}$ eV$^2$), respectively.

shown in the same figure are 95% C.L. excluded regions based on the present day-night and energy spectrum data. The excluded regions of neutrino oscillation parameters were estimated by a $\chi^2$ method which is defined as

$$\chi^2 = \sum_{day,night} \sum_{spectrum} \left( \frac{\phi_{Data} - \alpha \times \phi_{MC}(\sin^2 2\theta, \Delta m^2, \epsilon_j)}{\sigma} \right)^2$$

$$+ \sum_j \left( \frac{\epsilon_j}{\sigma_j} \right)^2,$$

where $\phi$ is the flux, $\alpha$ is the absolute normalization factor (a free parameter), and $\epsilon_j$'s are the systematic uncertainties in the measurement and prediction. We find that the "small-mixing MSW" and "vacuum oscillation" solutions are excluded at 95% C.L. We comment, however, that the small-mixing MSW solution gives a very good fit to the flux data from the six experiments. Therefore, if a combined analysis based on the flux, day-night and spectrum measurements is carried out, the small-mixing MSW solution is still allowed at 95% C.L.

In Figure 4 (left), neutrino oscillations between active neutrinos are assumed. However, there is a possibility that neutrino oscillations are between active and sterile neutrinos, $\nu_e \rightarrow \nu_{sterile}$, where $\nu_{sterile}$ is a hypothetical neutrino-like particle which does not interact with matter. Since the sterile neutrinos could have significant impact on both particle physics and cosmology, neutrino oscillations involving sterile neutrinos should be seriously studied. In Figure 4 (right), black regions show allowed neutrino oscillation parameters

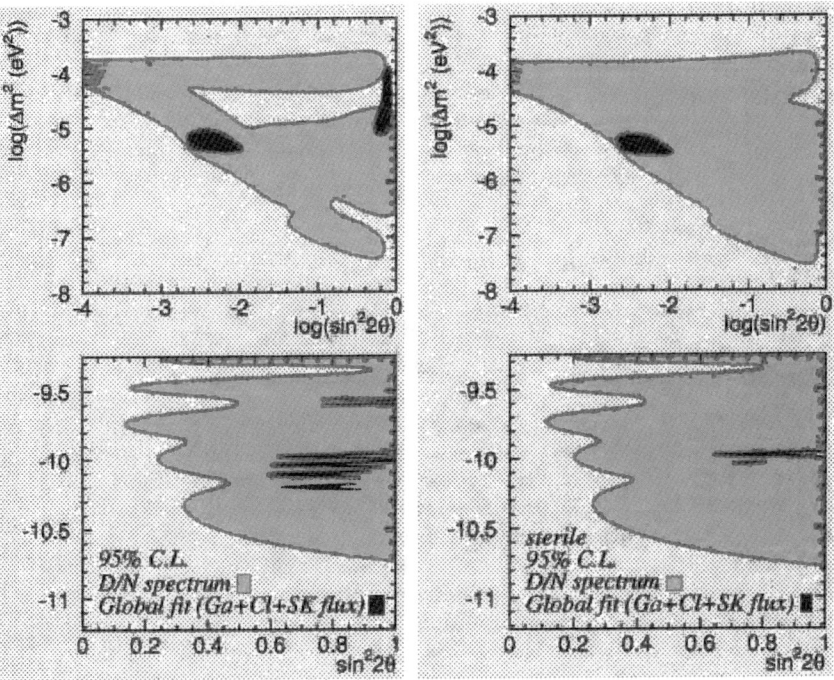

Figure 4. Black regions show 95%C.L. allowed regions of neutrino oscillation parameters based on the flux measurements from the Homestake, SAGE, Gallex, GNO and Super-Kamiokande. Shaded regions show 95%C.L. excluded regions from the day-night and spectrum measurements by Super-Kamiokande. The left figures assume $\nu_e \rightarrow \nu_x$ oscillations, where $\nu_x$ is either $\nu_\mu$ or $\nu_\tau$, while the right figures assume $\nu_e \rightarrow \nu_{sterile}$ oscillations.

obtained by using the flux measurements from various solar neutrino experiments, assuming $\nu_e \to \nu_{sterile}$ oscillations. Because of the much smaller Data/SSM value from the Homestake experiment compared with the Super-Kamiokande value, only oscillation parameters which show a large oscillation probability for the $^7$Be neutrinos are allowed. Also shown in the same figure are the excluded regions based on the present day-night and energy spectrum data. We find that the allowed solutions involving sterile neutrinos are excluded at 95% C.L. Therefore, we conclude that there is an inconsistency in the explanation of the solar neutrino problem by $\nu_e \to \nu_{sterile}$ oscillations.

## 3. Atmospheric Neutrinos

Cosmic ray interactions in the atmosphere produce atmospheric neutrinos. The uncertainty in the absolute flux prediction is about 20%. However, the $(\nu_\mu/\nu_e)$ flux ratio is predicted with an accuracy of better than 5% over a wide energy range between 0.1 and 30 GeV. In addition, the up/down ratio of the flux is calculated to an accuracy of better than a few % and is approximately unity above a few GeV neutrino energies independent of the details of the calculation. In the case of neutrino oscillations, these ratios can be different from the predictions.

The measured values of the $(\mu/e)$ ratio (which is closely related to the $(\nu_\mu/\nu_e)$ flux ratio) by Kamiokande were significantly smaller than the Monte Carlo (MC) predictions.[11,12] Also a zenith-angle dependent deficit of $\mu$-like events was observed by Kamiokande[12] at high energies. These observations strongly suggested neutrino oscillations, and therefore Kamiokande estimated the allowed parameter regions of neutrino oscillations. Because of the relatively small statistics, both $\nu_\mu \to \nu_e$ and $\nu_\mu \to \nu_\tau$ oscillations were allowed. Recently, long baseline reactor experiments, CHOOZ[13] and Palo Verde,[14] excluded the $\nu_\mu \to \nu_e$ solution of the atmospheric neutrino problem. In 1998, Super-Kamiokande reported that the atmospheric neutrino data gave evidence for neutrino oscillations.[15] After 1998, the statistics of the Super-Kamiokande data have been increased by more than a factor of two: 9178 fully contained (FC) events and 665 partially-contained (PC) events have been observed in a 70 kton·year exposure.

For the analysis of FC events, only single-ring events were used. Single-ring events were identified as $e$-like or $\mu$-like. The threshold momentum for the analysis were 100 and 200 MeV/$c$ for $e$-like and $\mu$-like events, respectively. The FC events were separated into "sub-GeV" ($E_{vis} < 1330$ MeV) and "multi-GeV" ($E_{vis} > 1330$ MeV) samples, where $E_{vis}$ was defined to be the energy of an electron that would produce the observed amount of Cherenkov light. In a full MC simulation, 87% (95%) of sub-GeV $e$-like ($\mu$-like) events were $\nu_e$ ($\nu_\mu$) charged-current (CC) interactions and 84% (99%) of multi-GeV $e$-like ($\mu$-like) events were $\nu_e$ ($\nu_\mu$) CC interactions. PC events were estimated to be 97% $\nu_\mu$ CC interactions; hence, all PC events were classified as $\mu$-like.

Table 1 summarizes the atmospheric neutrino data and the $(\mu/e)_{data}/(\mu/e)_{MC}$ measurements from Super-Kamiokande. The observed values of this ratio were significantly smaller than unity

Table 1. Summary of the atmospheric neutrino data and the $(\mu/e)_{data}/$ $(\mu/e)_{MC}$ ratio measurements in Super-Kamiokande (70 kton·yr). In the Monte Carlo simulation, the atmospheric neutrino flux without neutrino oscillations calculated by Honda *et al.* (1995) is used.

|  |  | Data | MC |
|---|---|---|---|
| Sub-GeV | $e$-like | 2531 | 2402.6 |
|  | $\mu$-like | 2486 | 3620.9 |
|  | multi-ring | 1885 | 2321.5 |
|  | $(\mu/e)_{data}/(\mu/e)_{MC} = 0.65 \pm 0.02 \pm 0.05$ | | |
| Multi-GeV | $e$-like | 576 | 555.4 |
|  | $\mu$-like(FC+PC) | 1167(502+665) | 1683.9(738.7+945.1) |
|  | multi-ring(FC) | 1198 | 1470.1 |
|  | $(\mu/e)_{data}/(\mu/e)_{MC} = 0.67^{+0.04}_{-0.03} \pm 0.08$ | | |

and were consistent with the Kamiokande,[11,12] IMB (sub-GeV)[17] and Soudan-2[18] results.

The zenith angle distributions for the FC and PC events from Super-Kamiokande are shown in Figure 5. The $\mu$-like data, especially multi-GeV $\mu$-like data, exhibited a strong up-down asymmetry in zenith angle ($\Theta$) while no significant asymmetry was observed in the $e$-like data. At low energies, below $\sim$400 MeV/$c$, the lepton direction has little correlation with the neutrino direction. The correlation angle becomes smaller with increasing lepton momentum. Therefore, the zenith angle dependence of the flux as a consequence of neutrino oscillations is largely washed out below 400 MeV/$c$ lepton momentum. With increasing momentum, the effect can be seen more clearly. We define the up-down double-ratio, $(U/D)_{Data}/(U/D)_{MC}$, where $U$ is the number of upward-going events ($-1 < \cos\Theta < -0.2$) and $D$ is the number of downward-going events ($0.2 < \cos\Theta < 1$). This ratio is expected to be consistent with unity within errors for no neutrino oscillations. The $(U/D)_{Data}/(U/D)_{MC}$ value for the multi-GeV FC+PC $\mu$-like events, $0.52 \pm 0.04$(stat.) $\pm 0.01$(syst.), deviated from unity by 9 standard deviations. This result, which is close to 0.5, suggest a near-maximal neutrino mixing, since $(U/D)_{Data}/(U/D)_{MC} \simeq 1 - (\sin^2 2\theta)/2$ to first approximation, assuming that there is no (full) oscillation effect for downward-(upward-)going particles, and averaging over the energy spectrum.

Energetic atmospheric $\nu_\mu$'s passing through the Earth interact with rock surrounding a detector and produce muons via CC interactions. These neutrino events are observed as upward going muons. Super-Kamiokande observed 1269 upward through-going muon events (including $9.1 \pm 0.8$ estimated background events) during 1138 detector live days.[19] In addition, 311 upward stopping muon events (including $21.4 \pm 8.8$ estimated background events) were observed during 1117 detector live days.[19] Figure 6 shows the zenith-angle distributions of the upward going muon fluxes.

The background events are expected to be in the most horizontal bins and are subtracted. The prediction for the through-going muons had a flatter zenith-angle distribution than the data. Similar

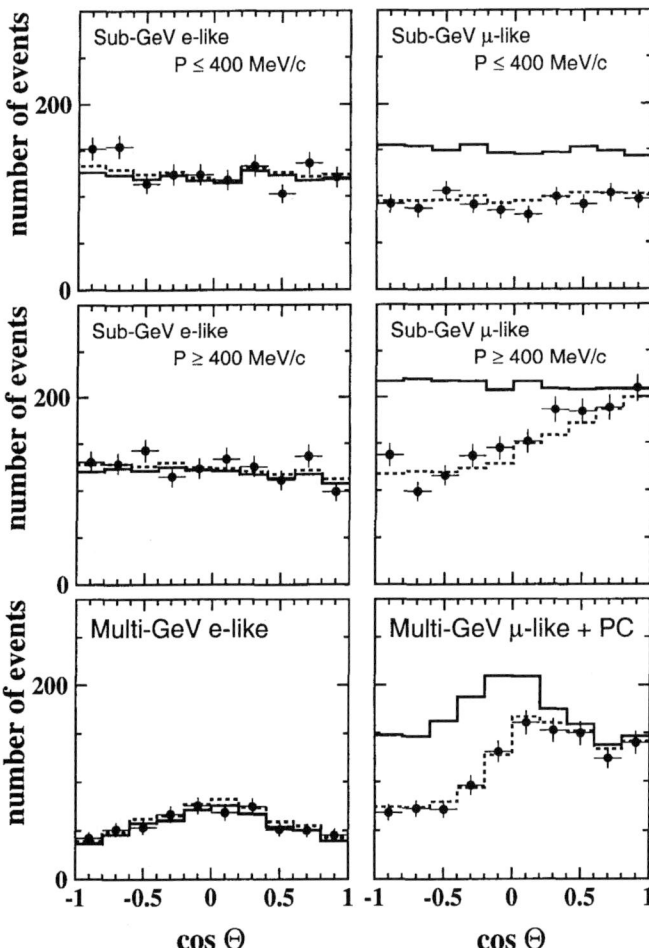

Figure 5. Zenith angle distributions for (left) $e$-like and (right) $\mu$-like events observed in Super-Kamiokande (70 kton·yr). The top figures show those with momentum lower than 400 MeV/$c$, the middle one for sub-GeV events with momentum higher than 400 MeV/$c$, and the bottom ones for multi-GeV events. Cos $\Theta = 1$ means down-going particles. The solid histograms show the MC prediction without neutrino oscillations. The dashed histograms show the MC prediction for $\nu_\mu \rightarrow \nu_\tau$ oscillations with $\sin^2 2\theta = 1$ and $\Delta m^2 = 3.2 \times 10^{-3}$ eV$^2$. In the oscillation histograms, the absolute normalization was adjusted to get the minimum $\chi^2$.

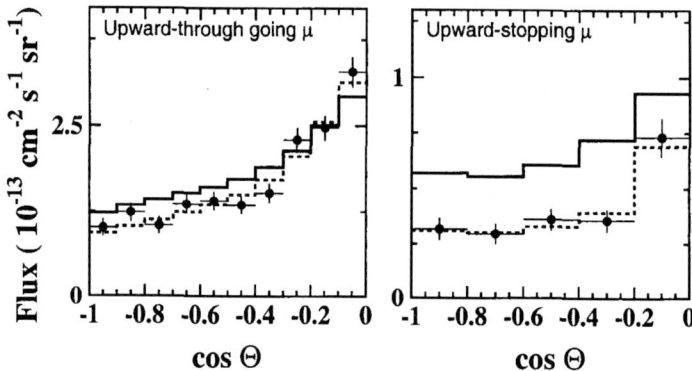

Figure 6. Zenith-angle distributions for upward-going muon fluxes observed in Super-Kamiokande. Circles with error bars show the upward through-going(left) and stopping(right) muon data and the histograms show the corresponding predictions. Error bars show statistical + experimental systematic errors. Estimated background events are subtracted. The solid histograms show the expected fluxes for the null neutrino oscillation case. The dashed histograms show the expected fluxes for the $\nu_\mu \to \nu_\tau$ oscillation case with $\sin^2 2\theta = 1.0$ and $\Delta m^2 = 3.2 \times 10^{-3}$ eV$^2$.

data were obtained in Kamiokande[20] and MACRO.[21] The observed flux of upward stopping muons by Super-Kamiokande was approximately a factor of two smaller than the prediction. These data are explained by neutrino oscillations.

### 3.1. *2 and 3 flavor neutrino oscillation analyses*

Since the contained events and upward-going muon events consistently suggested neutrino oscillations, an allowed region of the neutrino oscillation parameters assuming pure $\nu_\mu \to \nu_\tau$ oscillations was obtained using all the atmospheric neutrino data from Super-Kamiokande. In the oscillation analysis, various systematic errors were considered. They (and their size) are; absolute normalization (25%, but this was fitted as a free parameter), sub-GeV $(\mu/e)_{data}/(\mu/e)_{MC}$ (8%), multi-GeV $(\mu/e)_{data}/(\mu/e)_{MC}$ (12%), normalization of PC events relative to FC events (8%), normalization of upward stopping muons relative to FC+PC events (7%), normalization of upward through-going muons relative to upward-stopping muons (7%), energy spectrum index of neutrinos (0.05), sub-GeV up-down ratio (2.4%), multi-GeV up-down ratio (2.7%), vertical-horizontal ratio for FC+PC (4%) vertical-horizontal ratio for upward going muons (3%) and L/E (15%). The allowed region is shown in Figure 7 together with the allowed regions from Kamiokande,[20] and Soudan-2.[18] The best fit point was found at $(\sin^2 2\theta, \Delta m^2) = (1.0, 3.2 \times 10^{-3} \text{ eV}^2)$. Even if the analysis was extended to the unphysical region of $\sin^2 2\theta > 1$, the best fit point was the same. The 90% C.L. allowed region of the oscillation parameters was; $1.5 \times 10^{-3} < \Delta m^2 < 5 \times 10^{-3} \text{ eV}^2$ and $\sin^2 2\theta > 0.88$.

There is now intense flux calculation activity in which the fluxes are calculated based on a full three dimensional simulation of the particle trajectory and the Earth.[22,23] As the full three-dimensional calculations predict an enhancement of horizontal fluxes in the sub-GeV energy region, we have estimated the oscillation parameters based on fluxes by a three-dimensional calculation.[23] In spite

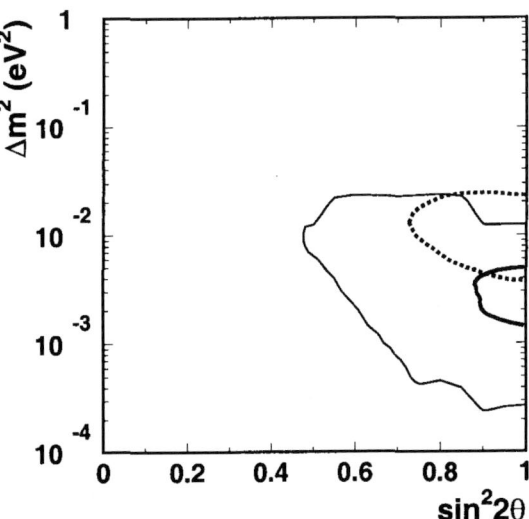

Figure 7. 90% C.L. allowed parameter region for $\nu_\mu \to \nu_\tau$ neutrino oscillations by a combined analysis of FC, PC, upward stopping muon and upward through-going muon events from Super-Kamiokande (thick line). Results from Kamiokande (FC + PC + upward-going muons, thick dashed line) and Soudan-2 (FC events, thin line) are also shown.

of the horizontal enhancement of the flux, the predicted zenith-angle distributions of $\mu$-like and $e$-like events are similar due to a large neutrino-lepton scattering angle in the sub-GeV energy region. Therefore, the allowed parameter regions based on one-dimensional and three-dimensional fluxes turned out to be similar (preliminary). (It is still possible that fluxes based on three-dimensional calculations could suggest slightly higher $\Delta m^2$ value than the one-dimensional fluxes do.) In any case, we need more studies for the three-dimensional flux calculation effects.

As an extension of this analysis, Super-Kamiokande has carried out a 3 flavor neutrino oscillations analysis. It was assumed that the mass difference between the lightest and the second lightest neutrinos is very small and therefore the effect of the oscillation between these two neutrino eigenstates is invisible in atmospheric neutrinos. The allowed region was obtained on a three-dimensional space of $\sin^2 \theta_{13}$, $\sin^2 \theta_{23}$ and $\Delta m^2 (\equiv \Delta m^2_{13} = \Delta m^2_{23})$. An allowed region on the $\sin^2 \theta_{13}$ and $\sin^2 \theta_{23}$ plane is shown in Figure 8. (Note that

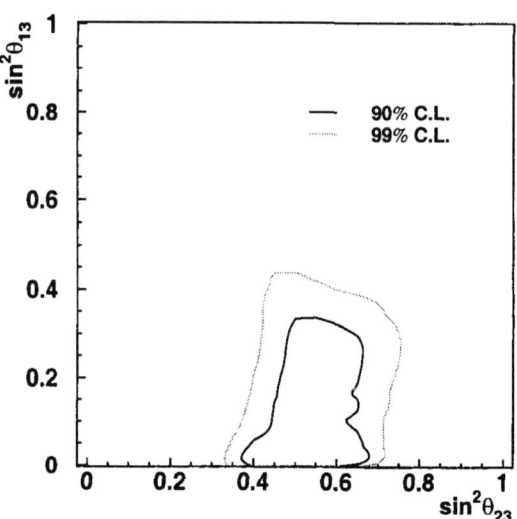

Figure 8. Allowed region on the $\sin^2 \theta_{13}$ and $\sin^2 \theta_{23}$ plane obtained by a 3 flavor analysis of the Super-Kamiokande FC+PC data. Black and gray lines show 90 and 99% C.L. allowed regions, respectively.

the axes are $\sin^2\theta_{ij}$, not $\sin^2 2\theta_{ij}$.) No evidence for non-zero $\sin^2\theta_{13}$ has been observed. This result is consistent with the CHOOZ[13] and Palo Verde[14] results. We comment that the constraint from the CHOOZ experiment[13] on $\sin^2\theta_{13}$ in the $\Delta m^2$ range between 1.5 and $5\times 10^{-3}$ eV$^2$ is much stronger that that from Super-Kamiokande.

## 3.2. *Energy dependence of neutrino oscillations*

There are models which predict neutrino oscillations with their energy dependence different from the standard neutrino oscillations generated by neutrino mass and mixing. (See, for example, Ref. 24 for the list of models.) Super-Kamiokande has studied the energy dependence of the neutrino oscillations. For this purpose, the atmospheric neutrino data (including upward going muons) were fitted with a form of $P(\nu_\mu \to \nu_\mu) = 1 - \sin^2\alpha \cdot \sin^2(\beta \cdot L \cdot E^n)$, where $\alpha, \beta$ and $n$ were fitted parameters.[24] Figure 9 shows the $n$ dependence of the $\chi^2$ value of the fit. The energy dependence

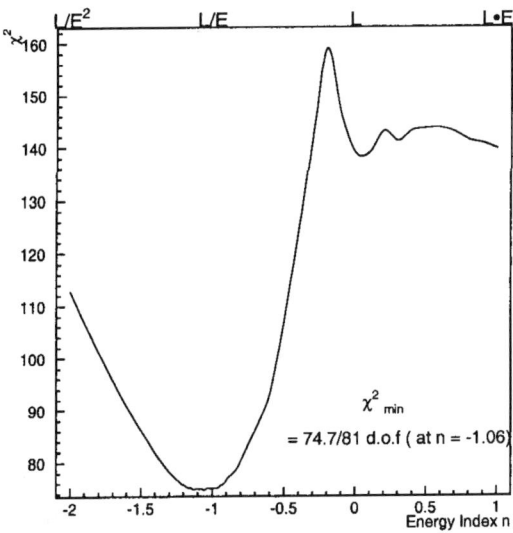

Figure 9. $\chi^2$ values of the fit to the atmospheric neutrino data as a function of the energy index $n$ and unconstrained oscillation factors $\alpha$ and $\beta$. The $\chi^2$ value is calculated assuming an oscillation probability $P(\nu_\mu \to \nu_\mu) = 1 - \sin^2\alpha \cdot \sin^2(\beta \cdot L \cdot E^n)$.

expected by the standard neutrino oscillation generated by neutrino mass and mixing ($n = -1$) was favored by the data (the fitted value of $n$ was $-1.06 \pm 0.14$), and models that predict other energy dependence of the oscillations were strongly disfavored.

### 3.3. $\nu_\mu \to \nu_\tau$ versus $\nu_\mu \to \nu_{sterile}$ oscillations

The atmospheric neutrino data are explained by $\nu_\mu \to \nu_\tau$ oscillations. However, since the expected event rate of $\nu_\tau$ interactions in the contained event sample is small (about 20 fully contained CC $\nu_\tau$ events are expected per 22.5 kton·year, which is less than 1% of the total atmospheric neutrino events), and since most of the neutral-current (NC) events are eliminated from the neutrino oscillation analysis that uses single-ring contained events, the data could also be explained by $\nu_\mu \to \nu_{sterile}$ neutrino oscillations. $\nu_{sterile}$ is a hypothetical neutrino-like particle which does not interact with matter by either CC or NC weak interactions. Indeed, according to the neutrino oscillation analysis using FC events, both $\nu_\mu \to \nu_\tau$ and $\nu_\mu \to \nu_{sterile}$ were almost equally allowed. Since the existence of $\nu_{sterile}$ has significant impact on particle physics and cosmology, it must be studied seriously whether $\nu_\mu \to \nu_{sterile}$ oscillations are really favored by the atmospheric neutrino data.

There are several ways to discriminate between the two possibilities. One possibility is to use a matter effect for upward going neutrino events.[25,26] In the case of $\nu_\mu \to \nu_\tau$ oscillations, the matter effect does not change the oscillation probability. On the other hand, in the case of $\nu_\mu \to \nu_{sterile}$ oscillations, the matter effect might change the oscillation probability significantly. For $\Delta m^2 \sim 3 \times 10^{-3}$ eV$^2$, the neutrino oscillation probability is expected to be significantly modified only for high energy ($> 20$ GeV) atmospheric neutrinos traveling through the Earth. In addition, the zenith angle distribution of a NC enriched sample is useful to discriminate between the two possibilities, because the NC events should be affected only for $\nu_\mu \to \nu_{sterile}$ oscillations.

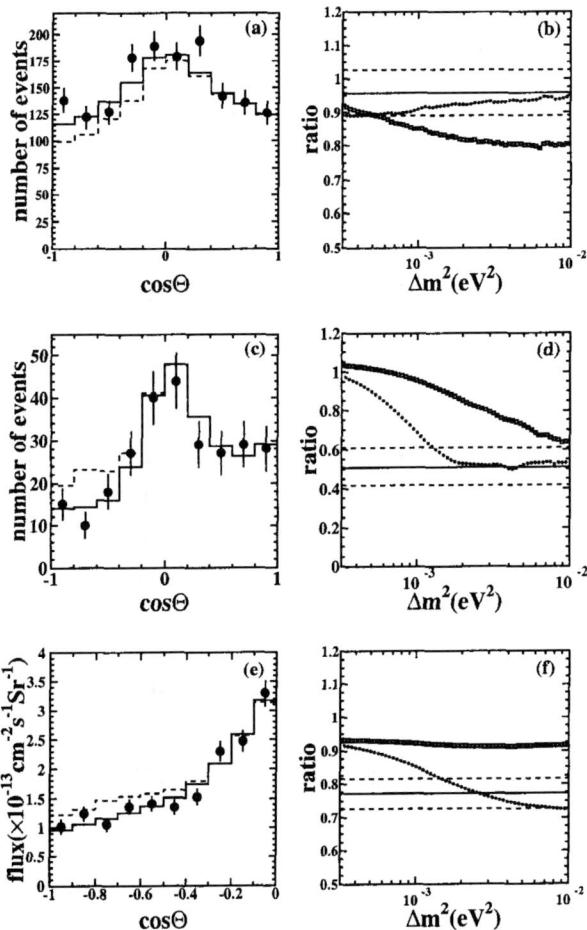

Figure 10. Zenith-angle distributions observed in Super-Kamiokande for: (a) multi-ring events with the most energetic ring being $e$-like and $E_{vis} > 400$ MeV, (c) PC events with $E_{vis} > 5$ GeV, and (e) upward through-going muon events. In (a), (c) and (e), solid (dashed) histograms show predictions for $\nu_\mu \rightarrow \nu_\tau$ ($\nu_\mu \rightarrow \nu_{sterile}$) oscillations with $\Delta m^2 = 3 \times 10^{-3}$ $eV^2$ and $\sin^2 2\theta = 1.0$. The predictions are normalized so that the number of observed and predicted events are equal at $0.4 < \cos\Theta < 1.0$ for (a) and (c) and at $-0.4 < \cos\Theta < 0.0$ for (e). In (b) and (d), expected $Up(-1 < \cos\Theta < -0.4)/Down(0.4 < \cos\Theta < 1)$ ratios of the corresponding data sets for $\nu_\mu \rightarrow \nu_\tau$ (dots) and $\nu_\mu \rightarrow \nu_{sterile}$ (empty squares) are plotted as a function of $\Delta m^2$. Also shown in the same figures are the $Up/Down$ ratios for the data. The solid lines show the central values for the data and the dashed lines show the $\pm 1\sigma$ statistical errors. In (f), expected $Vertical(-1 < \cos\Theta < -0.4)/Horizontal(-0.4 < \cos\Theta < 0)$ ratios for the upward through-going muon events for $\nu_\mu \rightarrow \nu_\tau$ (dots) and $\nu_\mu \rightarrow \nu_{sterile}$ (empty squares) are plotted as a function of $\Delta m^2$. Also shown in the same figure is the $Vertical/Horizontal$ ratio for the data. In these figures, $\sin^2 2\theta = 1$ is assumed for the expectations.

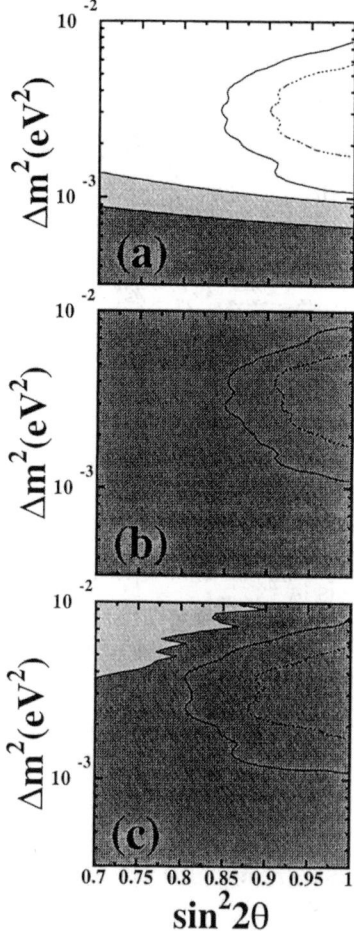

Figure 11. 90 and 99% C.L. allowed regions of neutrino oscillation parameters obtained by the FC events are shown by thin dotted and solid lines, respectively, for (a) $\nu_\mu \rightarrow \nu_\tau$, (b) $\nu_\mu \rightarrow \nu_{sterile}$ ($\Delta m^2 > 0$), and (c) $\nu_\mu \rightarrow \nu_{sterile}$ ($\Delta m^2 < 0$). Also shown are excluded regions from a combined analysis of the multi-ring, high-energy-PC and upward through going muon events. Light-grey and dark-grey regions show the 90% and 99% C.L. exclusion regions, respectively.

A result from Super-Kamiokande[27] was obtained from analyses of multi-ring, high-energy PC with $E_{vis} > 5$ GeV, and upward through-going muon events. Multi-ring events with the most energetic ring being $e$-like and $E_{vis} > 400$ MeV were used. The estimated fraction of NC events (in the absence of neutrino oscillations) was 29%. Figures 10 (a), (c) and (e) show the zenith angle distribution for these events, together with neutrino oscillation predictions. In all three samples, the differences between the two hypotheses are significant for upward-going directions. Since the shape of the zenith angle distributions is predicted accurately,[28] we studied the *Up/Down* or *Vertical/Horizontal* ratio depending on the data sample, as shown in Figures 10 (b), (d) and (f). The systematic errors in these ratios were estimated to be $(3-4)\%$ depending on the data sample.[27] In these three samples, the data disfavored $\nu_\mu \to \nu_{sterile}$ oscillations.

Finally, the two oscillation hypotheses were tested by combining these three studies by a $\chi^2$ method:

$$\chi^2 = \chi^2_{multi-ring} + \chi^2_{PC} + \chi^2_{upgoing-muon}.$$

If the above $\chi^2$ value is larger than 6.3 (11.3), the hypothesis is disfavored at 90 (99)% C.L. Figure 11 shows the results of this hypothesis test. Because the matter effect and therefore the neutrino oscillation probability depend on $(m_j^2 - m_i^2)$ rather than $|m_j^2 - m_i^2|$ for $\nu_\mu \to \nu_{sterile}$ oscillations, both positive and negative $\Delta m^2 (\equiv m_j^2 - m_i^2)$ were tested. The $\nu_\mu \to \nu_\tau$ oscillation hypothesis did not contradict the data in this analysis. On the other hand, the parameter regions suggested by the analysis of FC events for the pure $\nu_\mu \to \nu_{sterile}$ hypothesis were disfavored at 99% C.L.

## 4. Summary

Super-Kamiokande has been collecting solar and atmospheric neutrino data continuously since its start in 1996.

To distinguish between the solutions of the solar neutrino problem, the day-night effect and the energy spectrum of $^8$B solar neutrinos have been studied in detail. The present Super-Kamiokande data do not show any indication for the day-night effect nor the energy spectrum distortion that are expected for some neutrino oscillation parameters. The absence of these effects suggests that the small-mixing MSW and vacuum oscillations between $\nu_e \rightarrow \nu_{\mu \; or \; \tau}$ are disfavored. In addition, the $\nu_e \rightarrow \nu_{sterile}$ oscillation solutions, which had only two parameter regions at the small-mixing MSW and vacuum oscillation regions, are disfavored by the Super-Kamiokande energy spectrum data.

With the increased statistics of the atmospheric neutrino data, the main feature of the data has been essentially unchanged: Both the zenith angle distribution for $\mu$-like events and the $(\mu/e)$ values were significantly different from the predictions in the absence of neutrino oscillations. The data were in good agreement with $\nu_\mu \rightarrow \nu_\tau$ oscillations. This conclusion was supported by the upward-going muon data. By using all the atmospheric neutrino data from Super-Kamiokande, the 90% C.L. allowed region on $\nu_\mu \rightarrow \nu_\tau$ oscillation parameters was: $1.5 \times 10^{-3} < \Delta m^2 < 5 \times 10^{-3}$ eV$^2$ and $\sin^2 2\theta > 0.88$. The matter effect and the neutral current events have been used to discriminate between $\nu_\mu \rightarrow \nu_\tau$ and $\nu_\mu \rightarrow \nu_{sterile}$ oscillations. The data disfavored the $\nu_\mu \rightarrow \nu_{sterile}$ oscillation hypothesis at 99% C.L.

## Acknowledgments

This work was partly supported by the Japanese Ministry of Education, Science, Sports and Culture.

## References

1. B. T. Cleveland *et al.*, *Ap. J.* **496**, 505 (1998).
2. Y. Fukuda *et al.*, *Phys. Rev. Lett.* **77**, 1683 (1996).
3. J. N. Abdurashitov *et al.*, *Phys. Rev. Lett.* **83**, 4686 (1999).
4. W. Hampel *et al.*, *Phys. Lett. B* **447**, 127 (1999).
5. Y. Fukuda *et al.*, *Phys. Rev. Lett.* **81**, 1158 (1998); *Phys. Rev. Lett.* **82**, 1810 (1999); *Phys. Rev. Lett.* **82**, 2430 (1999).
6. M. Altmann *et al.*, *Phys. Lett. B* **490**, 16 (2000).
7. J. N. Bahcall, S. Basu and M. H. Pinsonneault, *Phys. Lett. B* **433**, 1 (1998).

8.  L. Wolfenstein, *Phys. Rev. D* **17**, 2369 (1978); S. P. Mikheyev and A. Yu. Smirnov, *Sov. J. Nucl. Phys.* **42**, 1441 (1985); *Nuovo Cimento* **C9**, 17 (1986).
9.  M. Nakahata *et al.*, *Nucl. Instrum. Methods A* **421**, 113 (1999).
10. E. Blaufuss *et al.*, hep-ex/0005014.
11. K. S. Hirata *et al.*, *Phys. Lett. B* **205**, 416 (1988); *Phys. Lett. B* **280**, 146 (1992).
12. Y. Fukuda *et al.*, *Phys. Lett. B* **335**, 237 (1994).
13. M. Apollonio *et al.*, *Phys. Lett. B* **420**, 397 (1998); *Phys. Lett. B* **466**, 415 (1999).
14. F. Boehm *et al.*, *Phys. Rev. Lett.* **84**, 3764 (2000); *Phys. Rev. D* **62**, 072002 (2000).
15. Y. Fukuda *et al.*, *Phys. Lett. B* **433**, 9 (1998); *Phys. Lett. B* **436**, 33 (1998); *Phys. Rev. Lett.* **81**, 1562 (1998).
16. M. Honda *et al.*, *Phys. Rev. D* **52**, 4985 (1995).
17. D. Casper *et al.*, *Phys. Rev. Lett.* **66**, 2561 (1991); R. Becker-Szendy *et al.*, *Phys. Rev. D* **46**, 3720 (1992).
18. W. W. M. Allison *et al.*, *Phys. Lett. B* **391**, 491 (1997); *Phys. Lett. B* **449**, 137 (1999); W. A. Mann, hep-ex/0007031.
19. Y. Fukuda *et al.*, *Phys. Rev. Lett.* **82**, 2644 (1999); *Phys. Lett. B* **467**, 185 (1999).
20. S. Hatakeyama *et al.*, *Phys. Rev. Lett.* **81**, 2016 (1998).
21. M. Ambrosio *et al.*, *Phys. Lett. B* **434**, 451 (1998); *Phys. Lett. B* **478**, 5 (2000).
22. G. Battistoni *et al.*, *Astropart. Phys.* **12**, 315 (2000).
23. P. Lipari, *Astropart. Phys.* **14**, 153 (2000).
24. G. L. Fogli *et al.*, *Phys. Rev. D* **60**, 053006 (1999).
25. Q. Y. Liu, S. P. Mikheyev, and A. Yu. Smirnov, *Phys. Lett. B* **440**, 319 (1998).
26. P. Lipari, and M. Lusignoli, *Phys. Rev. D* **58**, 073005 (1998).
27. S. Fukuda *et al.*, *Phys. Rev. Lett.* **85**, 3999 (2000).
28. P. Lipari, talk presented at the XIX International Conference on Neutrino Physics and Astrophysics, Sudbury, Canada, June 2000.

# Chapter 5

# Neutrino Oscillation Experiments: Super-Kamiokande, K2K and the JPARC Neutrino Project*

Takaaki Kajita

*Research Center for Cosmic Neutrinos, Institute for Cosmic Ray Research,
Univ. of Tokyo, Kashiwa-no-ha 5-1-5, Kashiwa, Chiba 277-8582, Japan
kajita@icrr.u-tokyo.ac.jp*

NEUTRINO oscillations have been studied using various neutrino sources including solar, atmospheric, reactor and accelerator neutrinos. Our understanding on neutrino masses and mixing angles has been improved significantly by recent experiments. The report mainly discusses the present status and the future prospect of our understanding of neutrino masses and mixing angles that are related to larger $\Delta m^2$.

## 1. Introduction

Neutrinos are known to be much lighter than any other quarks or charged leptons. Study of neutrino masses and mixing angles is one of a few ways to explore physics beyond the standard model of particle physics, since small neutrino masses are related to physics in very high energy scales.[1,2] Small neutrino masses can be studied by

---

*This article was originally published in *Compact Stars*, pp. 420–432 (2004).

neutrino flavor oscillations. For simplicity, we consider two-flavor neutrino oscillations. If neutrinos are massive, the flavor eigenstates, $\nu_\alpha$ and $\nu_\beta$, are expressed as combinations of the mass eigenstates, $\nu_i$ and $\nu_j$. The probability for a neutrino produced in a flavor state $\nu_\alpha$ to be observed in a flavor state $\nu_\beta$ after traveling a distance $L$ through the vacuum is:

$$P(\nu_\alpha \rightarrow \nu_\beta) = \sin^2 2\theta_{ij} \sin^2 \left( \frac{1.27\Delta m_{ij}^2(\text{ev}^2)L(\text{km})}{E_\nu(\text{GeV})} \right), \qquad (1)$$

where $E_\nu$ is the neutrino energy, $\theta_{ij}$ is the mixing angle between the flavor eigenstates and the mass eigenstates, and $\Delta m_{ij}^2 = m_{\nu j}^2 - m_{\nu i}^2$.

The above description has to be generalized to three-flavor oscillations. In the three-flavor oscillation framework, neutrino oscillations are parameterized by three mixing angles ($\theta_{12}$, $\theta_{23}$, and $\theta_{13}$), three mass squared differences ($\Delta m_{12}^2$, $\Delta m_{23}^2$, and $\Delta m_{13}^2$; among the three $\Delta m^2$'s, only two are independent) and one CP phase ($\delta$). However, if a neutrino mass hierarchy is assumed, the three $\Delta m^2$'s are approximated by two $\Delta m^2$'s, and neutrino oscillation lengths are significantly different for the two $\Delta m^2$'s. One $\Delta m^2(\Delta m_{12}^2)$ is related to solar neutrino experiments and the KamLAND reactor experiment. The other $\Delta m^2(\Delta m_{23}^2)$ is related to atmospheric, reactor and long baseline neutrino oscillation experiments. It is known that it is approximately correct to assume two-flavor oscillations for analyses of the present neutrino oscillation data. Therefore, in this article, we mostly discuss two flavor neutrino oscillations assuming two significantly different $\Delta m^2$'s. We mostly discuss experiments related to larger $\Delta m^2$. Especially, emphasis will be made on the atmospheric neutrino results from Super-Kamiokande, the K2K results and the sensitivity in JPARC-Kamioka neutrino project.

## 2. Present Data

### 2.1. *Atmospheric neutrino experiments*

The strongest evidence for $\nu_\mu \rightarrow \nu_\tau$ oscillation to date is given by atmospheric neutrino data from Super-Kamiokande.[3] The atmospheric neutrino flux is predicted to be up-down symmetric for the neutrino energies above a few GeV where the geomagnetic field effect can be neglected. On the other hand, neutrino oscillations with $\Delta m^2$ of about $3 \times 10^{-3}$ eV$^2$ predict a significant deficit of upward-going neutrino events. The first convincing evidence for oscillations was discovered by the zenith angle dependent deficit for muon neutrino events[3] (see also Ref. 4 for an earlier result.) Atmospheric neutrino experiments determine the $\nu_\mu \rightarrow \nu_\tau$ neutrino oscillation parameters utilizing the zenith angle and energy dependent deficit of muon neutrino events.

Recently, Super-Kamiokande has updated their neutrino interaction Monte Carlo simulation based on the K2K neutrino data. The detector Monte Carlo simulation and the event reconstruction have also been improved. In addition, a recent flux model based on a three dimensional calculation method[5] is used. Figure 1 shows the zenith angle distributions for various data samples from Super-Kamiokande. The zenith angle and energy dependent deficit of muon neutrino events is clearly seen. Consistent results have been obtained from Kamiokande,[6] Soudan-2[7] and MACRO.[8]

The allowed regions of $\nu_\mu \rightarrow \nu_\tau$ oscillation parameters from these experiments are shown in Figure 2. The allowed regions from various experiments are consistent. The best fit point from the Super-Kamiokande allowed regions is $2.0 \times 10^{-3}$ eV$^2$ for $\Delta m^2$ and 1.00 for $\sin^2 2\theta$ (preliminary). The 90% C.L. allowed region is $1.3 \times 10^{-3} < \Delta m^2 < 3.0 \times 10^{-3}$ eV$^2$ and $\sin^2 2\theta > 0.90$ (preliminary). The present best fit $\Delta m^2$ from Super-Kamiokande is lower than the previous estimate by about 20%. Each change in the flux model, interaction model, the detector simulation and the event reconstruction shifted $\Delta m^2$ in the same direction. As of this writing, Super-Kamiokande is finalizing the atmospheric neutrino analysis. The final results based on the data taken between 1996 and 2001 will be published soon.

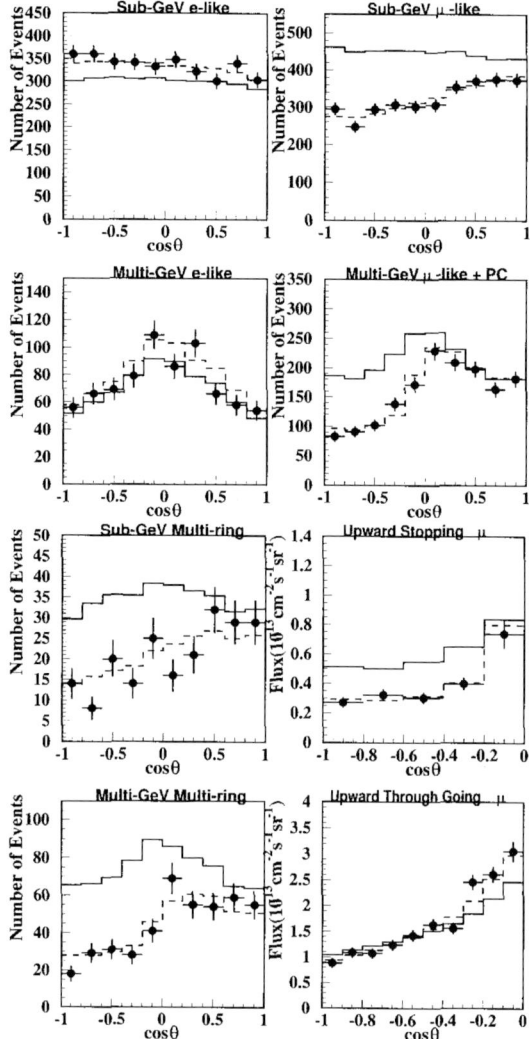

Figure 1.   Zenith angle distributions observed in Super-Kamiokande. The detector exposure is 1489 days (92 kton·yr) for fully-contained (FC) and partially-contained (PC) events, 1646 days for upward stopping muon and through going muon events. Events denoted by "sub-GeV" ("multi-GeV") have their visible energies lower than (higher than) 1.3 GeV. $\cos\Theta = 1(-1)$ means down-going (up-going). The solid histograms show the prediction without neutrino oscillations. The dotted histograms show the prediction with $\nu_\mu \to \nu_\tau$ oscillations ($\Delta m^2_{23} = 2.0 \times 10^{-3}$ eV$^2$, $\sin^2 2\theta_{23} = 1.0$). In the oscillation prediction, various uncertainty parameters such as the absolute normalization are adjusted to give the best fit to the data.

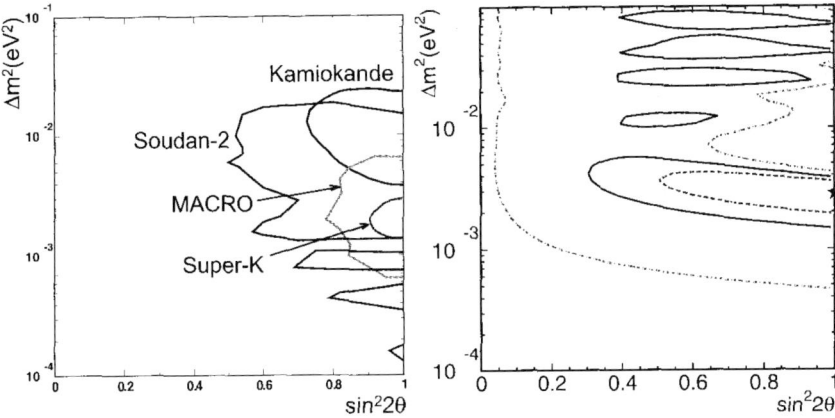

Figure 2. Allowed neutrino oscillation parameter regions for $\nu_\mu \rightarrow \nu_\tau$ from atmospheric neutrino experiments[6-8] at 90% C.L. (left) and the K2K[10] long baseline neutrino oscillation experiment at 68, 90 and 99% C.L. (right).

Several alternative hypotheses have been proposed to explain the atmospheric neutrino data. Most of them have been excluded or disfavored for various reasons. Neutrino oscillations between $\nu_\mu$ and $\nu_\tau$ give the best fit to the data. For maximal $\nu_\mu \rightarrow \nu_\tau$ oscillations with the $\Delta m^2$ preferred by the present data, it is expected that about 1 CC $\nu_\tau$ interaction should occur per kton per year. Super-Kamiokande has searched by CC $\nu_\tau$ interactions in the fully-contained atmospheric neutrino sample. Since the $\tau$ decays immediately after the production, a typical $\nu_\tau$ event looks like an energetic NC event in Super-Kamiokande. No single selection criterion can select the $\nu_\tau$ events efficiently. Therefore, maximum likelihood or neural network methods are used to maximize the detection sensitivity. Finally, the zenith angle distribution is used to statistically estimate the number of $\tau$ events, since only upward going events are expected for the $\tau$ events. Preliminary results showed that the data are consistent with the $\tau$ production.[9] However, the statistical significance is only 2 to 3 standard deviation level. More statistics and improved analysis are required for convincing evidence for $\tau$ production.

## 2.2. K2K

It is not trivial for atmospheric neutrino experiments to estimate the $\Delta m^2$ value precisely, since it is not possible to precisely estimate the $L_\nu / E_\nu$ value for each event. On the other hand, a long baseline experiment has only one neutrino flight length. Therefore, it is much easier for a long baseline neutrino oscillation experiment to estimate the $\Delta m^2$ value accurately.

K2K is the first long baseline neutrino oscillation experiment. Neutrinos are produced by using a 12 GeV proton beam at KEK. The neutrinos are detected in Super-Kamiokande. The neutrino flight length and the mean neutrino energy are 250 km and about 1.3 GeV, respectively. The estimated $\Delta m^2$ value from the atmospheric neutrino experiments suggests that the maximum oscillation effect should occur at the energies of less than 1 GeV for neutrinos whose flight length is 250 km. Therefore, K2K studies energy dependent deficit of muon neutrino events.

During the first data taking period between 1999 and 2001, $4.8 \times 10^{19}$ protons were delivered on target. 56 neutrino events have been observed in the far detector (Super-Kamiokande), while the expected number of events is $80.1^{+6.2}_{-5.4}$ for no oscillations.[10] In addition, K2K have studied the neutrino energy distribution using 29 single-ring $\mu$-like events. It is possible to calculate the neutrino energy from the muon energy and the direction assuming a quasi-elastic interation. Figure 3 shows the observed neutrino energy distribution from these single-ring $\mu$-like events. A deficit of events is observed between 0.5 and 1.0 GeV. (However, the statistics of the present data are not large enough to claim the evidence for energy dependent deficit.)

The allowed oscillation parameter region was estimated using the number of observed events and the reconstructed neutrino energy spectrum. The allowed parameter region is shown in Fig. 2. The best fit oscillation parameters were $\sin^2 2\theta = 1.00$ and $\Delta m^2 = 2.8 \times 10^{-3}$ eV$^2$. The allowed region from K2K is consistent with

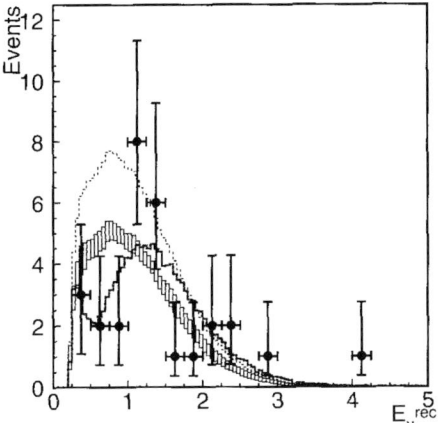

Figure 3. The reconstructed $E_\nu$ distribution for single-ring $\mu$-like events. Points with error bars show data. Box histogram shows expected spectrum without oscillations, where the height of the box show the systematic error. Solid line shows the best fit spectrum. These histograms are normalized by the number of observed events (29). Dashed line shows the expectation with no oscillations normalized to the expected number of events (44).

those from the atmospheric neutrino experiments. It should be noted that the allowed $\Delta m^2$ region from K2K is as small as that from Super-Kamiokande atmospheric neutrino data, while the statistics of the K2K data are less than 1% of the Super-Kamiokande atmospheric neutrino data. It was made possible, because the neutrino flight length is a single number in the K2K neutrino beam, while the flight length varies more than 3 orders of magnitude in the atmospheric neutrino beam.

K2K has resumed the experiment in Jan., 2003. Between Jan. and April, 2003, 16 additional events were observed in Super-Kamiokande, while the expected number of events was $26.4^{+2.3}_{-2.1}$. The event rate in the new data taking period is consistent with the previous rate.

A new front detector with full active scintillator trackers was installed and the data taking with this detector was started in Oct., 2003. This detector will make it possible to better understand the of low energy (of less than 1 GeV) neutrino interactions.

### 2.3. Limits on $\theta_{13}$

It is possible to get information on the other mixing angle ($\theta_{13}$) using the currently available data. Within an approximation that $\Delta m_{12}^2$ is much smaller than $\Delta m_{23}^2$ (i.e., $\Delta m_{12}^2 / \Delta m_{23}^2 = 0$), the neutrino oscillation probability can be written;

$$P(\nu_e \rightarrow \nu_e) = 1 - \sin^2 2\theta_{13} \sin^2 \left( \frac{1.27 \Delta m_{13}^2 (\text{eV}^2) L(\text{km})}{E_\nu (\text{GeV})} \right), \quad (2)$$

$$P(\nu_\mu \rightarrow \nu_e) = \sin^2 \theta_{23} \sin^2 2\theta_{13} \sin^2 \left( \frac{1.27 \Delta m_{13}^2 (\text{eV}^2) L(\text{km})}{E_\nu (\text{GeV})} \right). \quad (3)$$

The best limit is obtained by the CHOOZ[11] reactor experiment. In addition, constraints are obtained from Super-Kamiokande atmospheric neutrino experiment and the K2K experiment.[12] No evidence for finite $\theta_{13}$ has been observed. According to a combined analysis[13] of these results, the current upper limit on $\sin^2 \theta_{13}$ is 0.067 (at $3\sigma$). We note that the current limit on $\theta_{13}$ is a little weaker than the previous one due to the new $\Delta m^2$ value from Super-Kamiokande.

## 3. Future Neutrino Oscillation Experiments

Although the atmospheric neutrino data together with the K2K results, give convincing evidence for neutrino oscillations, there are several open questions: The measurement of $\Delta m_{23}^2$ by the atmospheric neutrino experiments is not very accurate. The observed effect has been the zenith angle and energy dependent deficit of CC $\nu_\mu$ events and the sinusoidal oscillation pattern has not been observed yet. The interactions of $\nu_\tau$, which must be generated by oscillations, have not been convincingly observed yet. Finally, $\theta_{13}$, the CP phase $\delta$ and the sign of $\Delta m^2$ are not known. Future long baseline neutrino oscillation experiments should address these issues and study further details of neutrino oscillations. Many of these experiments use accelerator generated neutrino beams. Since the typical neutrino energy is 1 GeV or higher, and since $\Delta m_{23}^2$ is about $2.5 \times 10^{-3}$ eV$^2$, the baseline length must be at least a few hundred km.

### 3.1. *MINOS and CNGS*

As of this writing, the MINOS experiment[14] is in the preparation stage. This experiment is able to study neutrino oscillations with much higher statistics than those in K2K. To produce neutrinos, the MINOS experiment uses the 120 GeV proton beam from the Main Injector at Fermilab. The MINOS detector is an iron-scintillator sampling calorimeter, which is located 730 km away from the target. The MINOS far detector has the total mass of 5.4 kton. The installation of the MINOS far detector has been completed in the summer of 2003. Since then, it has been taking atmospheric neutrino data. Because this detector is magnetized, it could give unique data on atmospheric neutrinos as well in the near future. The MINOS long baseline experiment is expected to start in early 2005. MINOS will improve our knowledge on $\Delta m^2$ significantly. Some improvement on the mixing angle measurements could also expected.

In Europe, CNGS (CERN Neutrino to Gran Sasso) project is in progress. 400 GeV proton beam from SPS at CERN produces high energy neutrinos whose mean energy is about 20–30 GeV. Neutrinos produced by this beam will be detected by detectors at Gran Sasso, which is 730 km away from the neutrino production point. OPERA[15] is an experiment for the CNGS project. This experiment is aimed to study neutrino oscillations by looking at the appearance of $\nu_\tau$ in the beam. The OPERA main detector consists of a 1.8 kton lead-emulsion target. Since it is not possible to scan all the images recorded in the emulsion, there are electronic detectors after each emulsion module in order to locate the interaction point in the emulsion. Only a portion of the emulsion that are located by the electronic detectors will be scanned to search for $\tau$ decay kinks. After

5 years of operation with $5 \times 10^{19}$ p.o.t. per year, the expected number of identified CC $\nu_\tau$ events are 4.3, 10.1 and 26.3 for $\Delta m_{23}^2 = 0.0016, 0.0025, 0.004\,eV^2$, while the expected background is 0.65. The experiment will start in 2006.

The other experiment for the CNGS project will be ICARUS.[16] It is a liquid argon TPC detector. The total mass will be 3 ktons. The excellent imaging capability of the detector should make it possible to carry out various neutrino oscillation studies. 600 ton ICARUS detector has been tested on the ground successfully.[17] The same detector will be installed at the Gran Sasso Laboratory soon. The sensitivity for $\nu_\tau$ appearance is very similar to that of OPERA. It has also a high sensitivity for the $\theta_{13}$ measurement.

### 3.2. *JPARC-Kamioka neutrino project*

There are as-yet-unobserved quantities related to neutrino oscillations: $\theta_{13}$, the sign of $\Delta m_{23}^2$ and the CP phase in the neutrino sector. These questions can be addressed by the next generation neutrino oscillation experiments. One possible experiment is the JPARC-Kamioka neutrino project.[18] JPARC is a high intensity proton accelerator complex that is under construction at JAERI, Tokai, Japan. 50 GeV PS will be used to produce the high intensity neutrino beam. The construction will be completed in the Japanese FY 2007. At the end of 2003 (about one month after this meeting) this neutrino project was approved by the Japanese government. The experiment will start in late 2008 or early 2009.

The main goals of the first phase of the project are the observation of non-zero $\sin^2 2\theta_{13}$ and the precise measurement of $\sin^2 2\theta_{23}$ and $\Delta m_{23}^2$. The 50 GeV PS is designed to deliver $3.3 \times 10^{14}$ protons every 3.4 seconds. The beam power is 0.75 MW. A future upgrade of the beam power to 4 MW is considered. The far detector is Super-Kamiokande. The baseline length of the experiment is 295 km. A 1 Mton water Cherenkov detector, Hyper-Kamiokande, is seriously considered as the far detector in the second phase of this neutrino project. The main goal of the second phase of the project is the observation of the CP violation effect.

A feature of this experiment is the use of a low-energy, narrow band, high-intensity neutrino beam. The neutrino energy will be tuned to the maximum oscillation energy. For $\Delta m_{23}^2 = 3.0 \times 10^{-3}$ eV$^2$, it is 715 MeV. To produce high intensity, narrow band beam, the off-axis beam technique will be used.[19] The axis of the beam is displaced by a few degrees from the far detector direction. Due to the two body decay kinematics of pions, the energy of neutrinos that pass through the far detector is low and almost independent of the pion energy spectrum. The neutrino energy can be adjusted by choosing the angle between the pion beam direction and the direction to the detector (off-axis angle).

Detailed Monte Carlo simulations have been carried out to estimate the expected neutrino spectrum and the number of events. If the off-axis angle is determined, the neutrino energy distribution is determined essentially. On the other hand, our present knowledge on $\Delta m_{23}^2$ is not precise enough to pre-determine the off-axis angle uniquely. Because of these conditions, the decay pipe is designed to accommodate off-axis angles between 2 and 3 degrees. The expected total numbers of events per year, which is equivalent to $10^{21}$ protons on target, are 3200 and 1100 for 2 and 3 degree off-axis beams, respectively. Throughout the discussion of the JPARC-Kamioka neurino project, we assume that the true $\Delta m_{23}^2$ is $3.0 \times 10^{-3}$ eV$^2$. We also assume the 2 degree off-axis beam.

The neutrino energy can be reconstructed accurately for quasi-elastic interactions assuming that the target nucleon is at rest. Figure 4 shows the reconstructed energy spectrum with and without neutrino oscillations assuming $\sin^2 2\theta_{23} = 1.0$ and $\Delta m_{23}^2 = 3.0 \times 10^{-3}$ eV$^2$. Events with single $\mu$-like Cherenkov ring are plotted. Since the peak of the neutrino energy distribution is tuned to the maximum oscillation energy, most of the muon neutrinos must be oscillated to tau neutrinos. From this figure it is possible to

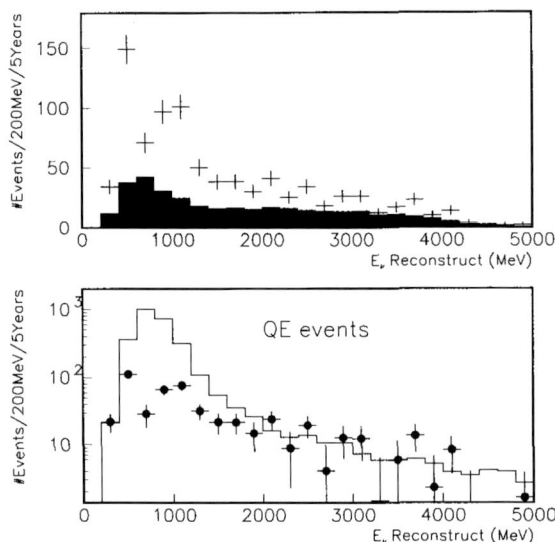

Figure 4.   Upper: Reconstructed neutrino energy distribution for 5 year operation of the JPARC-Kamioka experiment with the 2 degree off-axis beam.  Events with single $\mu$-like Cherenkov ring are plotted.  The shaded histogram shows the contribution of non-quasi-elastic events. $\sin^2 2\theta_{23} = 1.0$ and $\Delta m_{23}^2 = 3.0 \times 10^{-3}$ eV$^2$ are assumed.  The maximum oscillation effect is expected to occur at the 0.6 to 0.8 GeV energy bin. Lower: Reconstructed $E_\nu$ distributions after subtracting the non-quasi-elastic events.  The histogram and the dots show the non-oscillation and oscillation cases, respectively.

estimate the sensitivities in $\sin^2 2\theta_{23}$ and $\Delta m_{23}^2$.  The expected accuracy is 1% in $\sin^2 2\theta_{23}$ and $1 \times 10^{-4}$ eV$^2$ in $\Delta m_{23}^2$.  As far as $\Delta m_{23}^2$ is concerned, the sensitivity is limited by systematic uncertainties, such as uncertainties in absolute energy calibration or in nuclear effects that affect lepton energies.

The JPARC neurino beam has a small $\nu_e$ contamination (0.2% at the energy of the peak flux).  Furthermore, the $\nu_e$ appearance signal is maximized by tuning the neutrino energy at its oscillation maximum.  Thus, this experiment has a high sensitivity to $\theta_{13}$.  The signal should be searched for in the single-ring $e$-like events.  The main background processes are contamination of $\nu_e$ in the beam, and NC (mostly $\pi^0$) events.  Special cuts have been developed to reject these NC events (See[18] for details).  Figure 5 (left) shows the expected en-

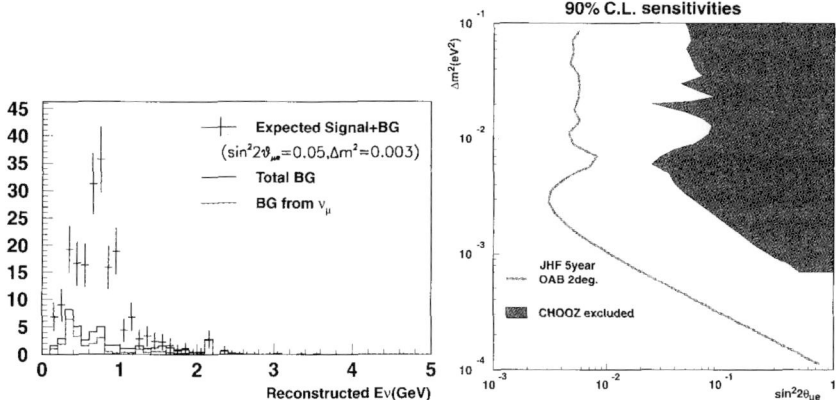

Figure 5. (Left) Expected electron appearance signal in the JPARC-Kamioka neutrino project. It is assumed that $\sin^2 2\theta_{13} = 0.1$ and 5 years of operation. (Right) 90% C.L. sensitivity in $\sin^2 \theta_{\mu e}$ ($\equiv 0.5 \cdot \sin^2 2\theta_{13}$) after 5 years of operation. Present limit from reactor experiment[11] is shown by a gray region. In these figures, effects due to the CP violation and the matter are neglected.

ergy distribution for the signal and background. Figure 5 (right) shows the expected sensitivity in $\sin^2 2\theta_{\mu e}$ ($\equiv \sin^2 \theta_{23} \cdot \sin^2 2\theta_{13}$ and assumed to be $0.5 \times \sin^2 2\theta_{13}$). The sensitivity of this experiment to $\sin^2 2\theta_{13}$ is about 0.006.

### 3.3. Competing projects

The neutrino beam produced by the Main Injector for the MINOS experiment can be used for an off-axis experiment. If a detector is located about 10 or 20 km away from the beam center, the mean beam energy can be about 2 and 1 GeV, respectively. As a far detector, about 50 kton, low-Z detector is considered. This experiment will have a slightly better sensitivity in $\sin^2 2\theta_{13}$ than the JPARC-Kamioka neutrino project.[23] Intensive R&Ds for the detector are in progress.

It is also possible to study $\theta_{13}$ using $\bar{\nu}_e$ from nuclear power reactors. Unlike accelerator experiments, the disappearance of $\bar{\nu}_e$ is almost a pure measurement of $\theta_{13}$. Detailed studies have shown that it must be possible to improve the $\sin^2 \theta_{13}$ limit by a factor of 5 to 10 (or more) with the present technology, if the systematic error is controlled to an accuracy of about 1%. (See, for example.[24])

### 3.4. *Toward the measurement of the CP violation*

There are now intense studies of designing experiments that detect CP violation effect. CP violation in the neutrino sector is considered to be the key to understand the baryon asymmetry of the Universe.[25] The CP violating phase is currently unknown and is expected to be observable if $\Delta m_{12}^2$ and $\Delta m_{23}^2$ are close enough so that both come into play in a single measurement and if the Jarlskog factor[26] $J = c_{12}c_{13}^2c_{23}s_{12}s_{13}s_{23}(\sin\delta)$ is large enough. Recent solar neutrino[20,21] and KamLAND[22] data confirm that $\theta_{12}$ and $\Delta m_{12}^2$ are large enough.

One example for such experiments is the second stage of the JPARC-Kamioka neutrino project. The CP violation phase can be measured by observing the difference in the neutrino oscillation probabilities between $\nu_\mu \rightarrow \nu_e$ and $\bar{\nu}_\mu \rightarrow \bar{\nu}_e$. To observe the CP violation effect, a huge detector (1 Mton Hyper-Kamiokande detector), a very high intensity proton accelerator (4 MW beam power) and about 8 years of operation will be required.[18]

Even more ambitious projects are under serious discussion and R&D. These discussions assume a very high intensity neutrino beam produced by muon storage rings (neutrino factories), see for example Ref. 27. The basic idea is to build a muon storage ring with a long straight section where muons decay to produce a collimated neutrino beam. The muon decay is well understood and yields a very well defined beam spectrum. A $\mu^+$ decays to $e^+$, $\bar{\nu}_\mu$ and $\nu_e$. Therefore, if a $\mu^-$ is observed in a neutrino interaction, this event must be an oscillation signal (or a background due to an imperfect detector resolution). Therefore, experiments with muon storage rings are expected to be very sensitive to a very small $\theta_{13}$ value. In addition, since the storage of positive and negative muons are possible, it is possible to study the CP violation effect. Detailed studies have shown that the sensitivity of these experiments to the CP violation effect is high. Also, since the baseline is very long (typically longer than 1000 km), the sign of $\Delta m_{23}^2$ can be determined easily using the matter effect.

## 4. Summary

Data from various neutrino oscillation experiments to date already give fairly detailed information on the neutrino masses and mixing angles. Two mixing angles, $\theta_{12}$ and $\theta_{23}$ are large. $\Delta m_{23}^2$ is significantly larger than $\Delta m_{12}^2$. However, our understanding of the neutrino masses and mixing angles is not complete. We do not know how small $\theta_{13}$ is. We have no knowledge about the CP phase in the neutrino sector. Future long baseline experiments should address many of these important questions. It is likely that future neutrino oscillation experiments will continue to contribute to our understanding of the particle physics.

This work has been supported by the Japanese Ministry of Education, Culture, Sports, Science and Technology.

## References

1. T. Yanagida, in Proc. of the Workshop on the Unified Theory and Baryon Number in the Universe (KEK Report No. 79-18) p. 95 (1979).
2. M. Gell-Mann, P. Ramond, R. Slansky, in Supergravity, p. 315 (1979).
3. Y. Fukuda *et al.*, *Phys. Rev. Lett.* **81**, 1562 (1998).
4. Y. Fukuda *et al.*, *Phys. Lett. B* **335**, 237 (1994).
5. M. Honda *et al.*, Proc. of the 28th International Cosmic Ray Conferences (ICRC2003), Tsukuba, Japan, 31 Jul–7 Aug 2003, Vol. 3, p. 1415.
6. S. Hatakeyama *et al.*, *Phys. Rev. Lett.* **81**, 2016 (1998).
7. M. Sanchez *et al.*, *Phys. Rev. D* **68**, 113004 (2003).
8. M. Ambrosio *et al.*, *Phys. Lett. B* **566**, 35 (2003).
9. C. Saji, for the Super-Kamiokande collaboration, in Proc. of the 28th International Cosmic Ray Conference, Tsukuba, Japan, July-Aug. 2003, Vol. 3, p. 1267.
10. M. H. Ahn *et al.*, *Phys. Rev. Lett.* **90**, 041801 (2003).
11. M. Apollonio *et al.*, *Phys. Lett. B* **466**, 415 (1999).
12. K. Nishikawa, talk presented at the XXI International Symposium on Lepton and Photon Interactions at High Energies, Fermilab, USA, Aug. 2003.

13. G. L. Fogli *et al.*, hep-ph/0308055.
14. D. Michael, *Nucl. Phys. B (Proc. Suppl.)* **118**, 189 (2003).
15. OPERA collaboration, M. Guler *et al.*, CERN-SPSC-P-318, LNGS-P25-00 (2000).
16. A. Rubbia, *Nucl. Phys. B (Proc. Suppl.)* **91**, 223 (2001).
17. O. Palamara, for the ICARUS Collaboration, *Nucl. Phys. B (Proc. Suppl.)* **110**, 329 (2002).
18. Y. Itow *et al.*, hep-ex/0106019.
19. D. Beavis *et al.*, Proposal, BNL AGS E-889 (1995).
20. M. B. Smy *et al.*, hep-ex/0309011.
21. S. N. Ahmed *et al.*, nucl-ex/0309004.
22. K. Eguchi *et al.*, *Phys. Rev. Lett.* **90**, 021802 (2003).
23. D. Ayres *et al.*, hep-ex/0210005.
24. H. Minakata *et al.*, *Phys. Rev. D* **68**, 033017 (2003): P. Huber, *et al.*, *Nucl. Phys. B* **665**, 487 (2003).
25. M. Fukugita and T. Yanagida, *Phys. Lett. B* **174**, 45 (1986).
26. C. Jariskog, *Phys. Rev. Lett.* **55**, 1039 (1985).
27. S. Geer, *J. Phys. G: Nucl. Part. Phys.* **29**, 1485 (2003).

# Chapter 6

# Present and Future Neutrino Oscillation Experiments*

Takaaki Kajita

*Research Center for Cosmic Neutrinos, Institute for Cosmic Ray Research,
Univ. of Tokyo, Kashiwa-no-ha 5-1-5, Kashiwa, Chiba 277-8582, Japan
kajita@icrr.u-tokyo.ac.jp*

NEUTRINO oscillations have been studied using various neutrino sources including solar, atmospheric, reactor and accelerator neutrinos. Our understanding on neutrino masses and mixing angles has been improved significantly by recent experiments. This report summarizes the present status and the future prospect of our understanding of neutrino masses and mixing angles.

## 1. Introduction

Neutrinos are known to be much lighter than any other quarks or charged leptons. Study of neutrino masses and mixing angles is one of a few ways to explore physics beyond the standard model of particle physics.[1,2] Small neutrino masses can be studied by neutrino flavor oscillations. For simplicity, we consider two-flavor neutrino oscillations. If neutrinos are massive, the flavor eigenstates, $\nu_\alpha$ and $\nu_\beta$, are expressed as combinations of the mass eigenstates, $\nu_i$ and $\nu_j$.

---

*This article was originally published in *Thinking, Observing and Mining the Universe*, pp. 199–208 (2004).

The probability for a neutrino produced in a flavor state $\nu_\alpha$ to be observed in a flavor state $\nu_\beta$ after traveling a distance $L$ through the vacuum is:

$$P(\nu_\alpha \rightarrow \nu_\beta) = \sin^2 2\theta_{ij} \sin^2 \left( \frac{1.27 \Delta m_{ij}^2 (\text{eV}^2) L(\text{km})}{E_\nu (\text{GeV})} \right), \qquad (1)$$

where $E_\nu$ is the neutrino energy, $\theta_{ij}$ is the mixing angle between the flavor eigenstates and the mass eigenstates, and $\Delta m_{ij}^2 = m_{\nu j}^2 - m_{\nu i}^2$.

The above description has to be generalized to three-flavor oscillations. In the three-flavor oscillation framework, neutrino oscillations are parameterized by three mixing angles ($\theta_{12}$, $\theta_{23}$, and $\theta_{13}$), three mass squared differences ($\Delta m_{12}^2$, $\Delta m_{23}^2$, and $\Delta m_{13}^2$; among the three $\Delta m^2$'s, only two are independent) and one CP phase ($\delta$). However, if a neutrino mass hierarchy is assumed, the three $\Delta m^2$'s are approximated by two $\Delta m^2$'s, and neutrino oscillation lengths are significantly different for the two $\Delta m^2$'s. One $\Delta m^2 (\Delta m_{12}^2)$ is related to solar neutrino experiments and the KamLAND reactor experiment. The other $\Delta m^2 (\Delta m_{23}^2)$ is related to atmospheric, reactor and long baseline neutrino oscillation experiments. It is known that it is approximately correct to assume two-flavor oscillations for analyses of the present neutrino oscillation data. Therefore, in this article, we mostly discuss two flavor neutrino oscillations assuming two significantly different $\Delta m^2$'s.

## 2. Solar and KamLAND Experiments

The observed fluxes from various solar neutrino experiments have been significantly lower than the Standard Solar Model (SSM) predictions. It has been known that the solar neutrino data cannot be explained by reasonable modifications of the solar models. On the other hand, it has been known that neutrino oscillations naturally explain the solar neutrino data. Several allowed oscillation parameter regions have been identified.

The SNO detector is a 1 kton heavy water Cherenkov detector and was designed to confirm the solar neutrino oscillations inde-

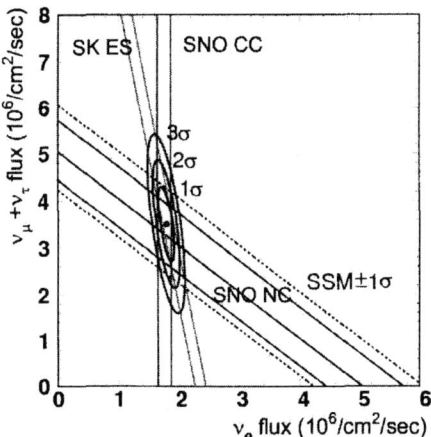

Figure 1. The observed $v_e$ and $v_{\mu \text{ or } \tau}$ fluxes from SNO by CC and NC interactions. The NC events are detected through $v_x d \rightarrow v_x pn$, $nd \rightarrow t\gamma$ (6 MeV) (pure D$_2$O phase). The constraint from Super-Kamiokande measurement by neutrino electron scattering is also shown. This figure is modified from Fig. 3 of Ref. 3.

pendent of the details of the solar model predictions. It can detect $^8$B solar neutrinos by both charged-current (CC; $v_e d \rightarrow e^+ pp$) and neutral-current (NC; $v_x d \rightarrow v_x pn$) interactions off deuterium. The CC interaction only measures the flux of $v_e$. On the other hand, the NC interaction measures the ($v_e + v_\mu + v_\tau$) flux. The observed CC/NC ratio compared with the expected ratio should give the direct test for oscillations. In the first phase of the SNO experiment, the NC signal was observed by detecting a 6 MeV gamma ray that was generated by $nd \rightarrow t\gamma$. Figure 1 shows the solar $v_e$ and $v_{\mu \text{ or } \tau}$ fluxes observed in the first phase of SNO.[3] The constraint from Super-Kamiokande[4] using neutrino electron scattering is also shown. It is clear that the data give evidence for non-zero $v_{\mu \text{ or } \tau}$ flux and therefore evidence for neutrino flavor change (oscillations). In addition, the existence of an overlapping region in this figure suggest the consistency between data from SNO and Super-Kamiokande.

The data are used to constrain the oscillation parameter regions. To constrain the parameter regions, additional information such as day and night fluxes and energy spectrum from Super-Kamiokande[4] and SNO[3] are used. These experiments have not observed any significant day-night flux difference or the energy spectrum distortion. Only the LMA (Large Mixing Angle) solution remained. The other solutions were disfavored at 99% or higher C.L.

## 2.1. *KamLAND*

Since the solar neutrino experiments suggest the LMA solution, it is possible for KamLAND to directly test the solar neutrino solution. KamLAND is a long baseline reactor neutrino experiment with 1 kton of liquid scintillator. The mean neutrino energy is a few MeV and the mean neutrino flight length is about 180 km. Therefore, this experiment can prove $\Delta m^2$ as low as $10^{-5}$ eV$^2$ covering the entire LMA region. The first KamLAND result was presented in Dec. 2002.[5] In an detector exposure of 162 ton·yr (145.1 days) the ratio of the number of observed events to the expected number of events without disappearance was $0.611 \pm 0.085(\text{stat}) \pm 0.041(\text{syst})$ for $\bar{\nu}_e$ energies above 3.4 MeV. The deficit of events was inconsistent with the expected rate for no oscillation at the 99.95% C.L., thus confirming the LMA solution of the solar neutrino problem. In addition, the observed energy spectrum constrain the neutrino oscillation parameters, especially $\Delta m^2$. Figure 2 shows the observed energy specrum in KamLAND. The KamLAND result constrained the allowed $\Delta m^2$ region in the LMA region. However, the present KamLAND energy spectrum allows two solutions, LMA-I ($\Delta m^2$ being about $7 \times 10^{-5}$ eV$^2$) and LMA-II ($\Delta m^2$ being about $15 \times 10^{-5}$ eV$^2$).

## 2.2. *Recent solar neutrino data*

It is possible to separate the two solutions by looking at the day-night flux difference and the CC/NC ratio in solar neutrino

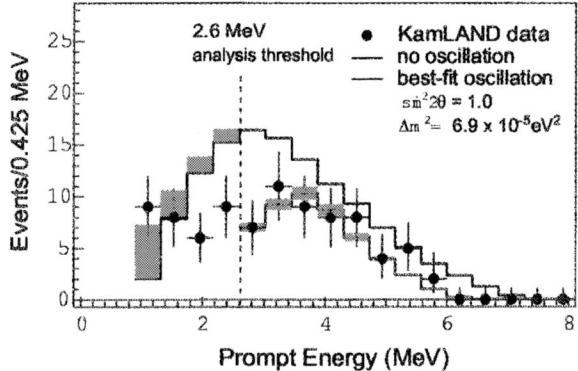

Figure 2. Energy spectrum of the observed reactor events in KamLAND[5] (solid circles with error bars), along with the expected no oscillation spectrum and best fit (lower histogram) including neutrino oscillations. The expected contribution of geo-neutrinos and background are also shown. The shaded band indicates the systematic error in the best-fit spectrum. The vertical dashed line corresponds to the analysis threshold at 2.6 MeV.

experiments.[6] Recent results from Super-Kamiokande[7] based on a maximum likelihood analysis of the solar neutrino data show that the observed day-night flux data are more consistent with the prediction for LMA-I than that for LMA-II, disfavoring the LMA-II solution at about 99% C.L.

Recently, the SNO collaboration has shown the initial data from the salt phase. 2 tons of $NaCl$ were added into 1 kton of heavy water. Neutrons that are produced by NC interactions are detected through $n^{35}Cl \rightarrow ^{36}Cl\gamma's$. Because of the much higher neutron capture cross section in $Cl$ and higher energy for the sum of the gamma rays, the efficiency for NC detection has been improved significantly. The observed CC/NC ratio was $0.306 \pm 0.026(\text{stat}) \pm 0.024(\text{syst})$.[8] This ratio is smaller than the prediction for the LMA-II region and disfavored the LMA-II region. Figure 3 shows the allowed region for neutrino oscillation parameters. Data from various solar neutrino experiments and the KamLAND experiment are used. Only the LMA-I region is allowed at 99% C.L. The maximum mixing angle is excluded at more than 5 standard deviations. Because of the matter effect for oscillations between $v_e$ and $v_x$, where

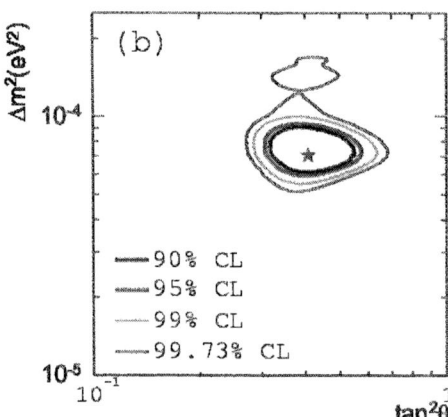

Figure 3.    Allowed neutrino oscillation parameter regions for $\nu_e \rightarrow \nu_{\mu \, or \, \tau}$ from a global analysis of the KamLAND and solar neutrino experiments.[8] The SNO salt data are included.

$x = \mu$ or $\tau$, the mixing angle smaller and larger than $45°$ can be distinguished, and therefore the horizontal axis shows $\tan^2 \theta_{12}$.

Our understanding for $\nu_e \rightarrow \nu_{\mu \, or \, \tau}$ oscillations has been improved significantly over the last several years. The allowed parameter regions are $\Delta m^2_{12} = 7.1^{+1.2}_{-0.6}$ and $\theta_{12} = 32.5^{+2.4}_{-2.3}$ degrees at 68 %C.L.[8]

### 2.3. Atmospheric and K2K experiments

The atmospheric neutrino flux is predicted to be up-down symmetric for the neutrinos above a few GeV where the geomagnetic field effect can be neglected. On the other hand, neutrino oscillations with $\Delta m^2$ of about $3 \times 10^{-3}$ predict that a significant deficit of upward-going neutrino events should be observed. The first convincing evidence for oscillations was discovered by the zenith angle dependent deficit for muon neutrino events[9] (see also Ref. 10 for an earlier result.) Atmospheric neutrino experiments determine the $\nu_\mu \rightarrow \mu_\tau$ neutrino oscillation parameters utilizing the zenith angle and energy dependent deficit of muon neutrino events.

Recently, Super-Kamiokande has updated their neutrino interaction Monte Carlo simulation based on the K2K neutrino data.

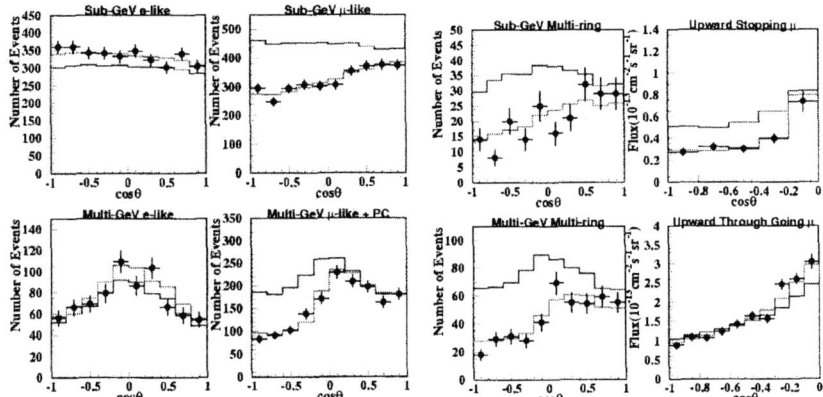

Figure 4. Zenith angle distributions observed in Super-Kamiokande. The detector exposure is 1489 days (92 kton·yr) for FC and PC events, 1646 days for upward stopping muon and through going muon events. $\cos\Theta = 1(-1)$ means down-going (up-going). The solid (or red) histograms show the prediction without neutrino oscillations. The dotted (or green) histograms show the prediction with $\nu_\mu \rightarrow \nu_\tau$ oscillations ($\Delta m_{23}^2 = 2.0 \times 10^{-3}$ eV$^2$, $\sin^2 2\theta_{23} = 1.0$). In the oscillation prediction, various uncertainty parameters such as the absolute normalization are adjusted to give the best fit to the data.

The detector Monte Carlo simulation and the event reconstruction have also been improved. In addition, a recent flux model based on a three dimensional calculation method is used. Figure 4 shows the zenith angle distributions for various data samples from Super-Kamiokande. The zenith angle and energy dependent deficit of muon neutrino events is clearly seen. Consistent results have been obtained from Kamiokande,[10,11] Soudan-2[12] and MACRO.[13]

The allowed regions of $\nu_\mu \rightarrow \nu_\tau$ oscillation parameters from these experiments are shown in Figure 5. The allowed regions from various experiments are consistent. The best fit point from the Super-Kamiokande allowed regions is $2.0 \times 10^{-3}$ eV$^2$ for $\Delta m^2$ and 1.00 for $\sin^2 2\theta$ (preliminary). The 90% C.L. allowed region is $1.3 \times 10^{-3} < \Delta m^2 < 3.0 \times 10^{-3}$ eV$^2$ and $\sin^2 2\theta > 0.90$ (preliminary). The present best fit $\Delta m^2$ from Super-Kamiokande is lower than the previous estimate by about 20%. Accidentally, each change in the flux model, interaction model, the detector simulation and the event reconstruction shifted $\Delta m^2$ in the same direction.

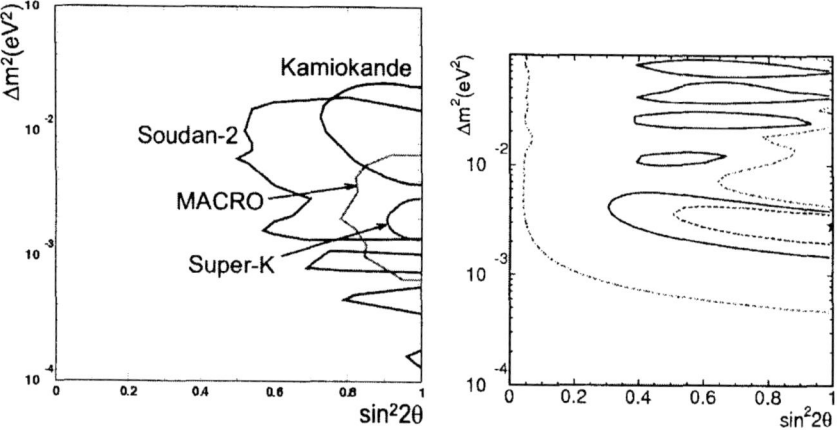

Figure 5. Allowed neutrino oscillation parameter regions for $\nu_\mu \rightarrow \nu_\tau$ from atmospheric neutrino experiments[11–13] (left) and the K2K[15] long baseline neutrino oscillation experiment (right).

Several alternative hypotheses have been proposed to explain the atmospheric neutrino data. Most of them have been excluded or disfavored for various reasons and neutrino oscillations between $\nu_\mu$ and $\nu_\tau$ give the best fit to the data. For maximal $\nu_\mu \rightarrow \nu_\tau$ oscillations with the $\Delta m^2$ preferred by the present data, it is expected that about 1 CC $\nu_\tau$ interaction should occur per kton per year. Super-Kamiokande has searched for CC $\nu_\tau$ interactions in the fully-contained atmospheric neutrino sample. Since the $\tau$ decays immediately after the production, a typical $\nu_\tau$ event looks like an energetic NC event in Super-Kamiokande. No single selection criterion can select the $\nu_\tau$ events efficiently. Therefore, maximum likelihood or neural network methods are used to maximize the detection sensitivity. Finally, the zenith angle distribution is used to statistically estimate the number of $\tau$ events, since only upward going events are expected for the $\tau$ events. Preliminary results showed that the data are consistent with the $\tau$ production.[14] However, the statistical significance is only 2 to 3 standard deviation level. More statistics and improved analysis are required for convincing evidence for $\tau$ production.

## 2.4. *K2K*

It is not trivial for atmospheric neutrino experiments to estimate the $\Delta m^2$ value precisely, since it is not possible to precisely estimate the $L_\nu / E_\nu$ value event by event. On the other hand, a long baseline experiments has only one neutrino flight length. Therefore, it is much easier for a long baseline neutrino oscillation experiment to estimate the $\Delta m^2$ value accurately.

K2K is the first long baseline neutrino oscillation experiment. Neutrinos are produced by using a 12 GeV proton beam at KEK. The neutrinos are detected in Super-Kamiokande. The neutrino flight length is 250 km. The mean neutrino energy is about 1.5 GeV. The estimated $\Delta m^2$ value from the atmospheric neutrino experiments suggests that the maximum oscillation effect should occur for neutrinos with the energies less than 1 GeV and the flight length of 250 km. Therefore, K2K studies energy dependent deficit of muon neutrino events.

During the first data taking period between 1999 and 2001, 56 neutrino events have been observed in the far detector (Super-Kamiokande), while the expected number of events is $80.1^{+6.2}_{-5.4}$ for no oscillations.[10] In addition, K2K have studied the neutrino energy distribution using 29 single-ring muon-like events. A deficit of events is observed between 0.5 and 1.0 GeV. (However, the statistics is too few to claim the evidence for energy dependent deficit with the present statistics.) The allowed oscillation parameter region was estimated using the number of observed events and the reconstructed neutrino energy spectrum. The allowed parameter region is shown in Figure 5. The allowed region from K2K is consistent with those from the atmospheric neutrino experiments. It should be noted that the allowed $\Delta m^2$ region from K2K is as small as that from Super-Kamiokande atmospheric neutrino data, while the statistics of the K2K data are less than 1% of the Super-Kamiokande atmospheric neutrino data.

### 2.5. *Limits on* $\theta_{13}$

It is possible to get information on the other mixing angle ($\theta_{13}$) using the presently available data. Within an approximation that $\Delta m_{12}^2$ is much lower than $\Delta m_{23}^2$ (i.e., $\Delta m_{12}^2/\Delta m_{23}^2 = 0$), the neutrino oscillation probability can be written;

$$P(\nu_e \to \nu_e) = 1 - \sin^2 2\theta_{13} \sin^2 \left( \frac{1.27\Delta m_{13}^2 (\text{eV}^2) L(\text{km})}{E_\nu(\text{GeV})} \right), \quad (2)$$

$$P(\nu_\mu \to \nu_e) = \sin^2 \theta_{23} \sin^2 2\theta_{13} \sin^2 \left( \frac{1.27\Delta m_{13}^2 (\text{eV}^2) L(\text{km})}{E_\nu(\text{GeV})} \right). \quad (3)$$

The best limit is obtained by the CHOOZ[16] reactor experiment. In addition, constraints are obtained from Super-Kamiokande atmospheric neutrino experiment and the K2K experiment.[17] No evidence for finite $\theta_{13}$ has been observed. According to a combined analysis[18] of these results, the current upper limit on $\sin^2 \theta_{13}$ is 0.067 (at $3\sigma$). We note that the current limit on $\theta_{13}$ is a little weaker than before due to the new $\Delta m^2$ value from Super-Kamiokande.

## 3. Future Neutrino Oscillation Experiments

Although the atmospheric neutrino data together with the K2K results, give convincing evidence for neutrino oscillations, there are several open questions: The measurement of $\Delta m_{23}^2$ by the atmospheric neutrino experiments is not very accurate. The observed effect has been the zenith angle and energy dependent deficit of CC $\nu_\mu$ events and the sinusoidal oscillation pattern has not been observed yet. The interactions of $\nu_\tau$ that must be generated by oscillations have not been convincingly observed yet. Finally, $\theta_{13}$, the CP phase $\delta$ and the sign of $\Delta m^2$ are not known. Future long baseline neutrino oscillation experiments should address these issues and study further details of neutrino oscillations. Many of these experiments use accelerator generated neutrino beams. Since the typical neutrino energy is 1 GeV or higher, and since $\Delta m_{23}^2$ is about $2.5 \times 10^{-3}$ eV$^2$, the baseline length must be at least a few hundred km.

As of this writing, the MINOS experiment[19] is in the preparation stage. This experiment is able to study neutrino oscillations with high statistics. To produce neutrinos, the MINOS experiment uses the 120 GeV proton beam from the Main Injector at Fermilab. The MINOS detector is an iron-scintillator sampling calorimeter, which is located 730 km away from the target. The MINOS far detector has the total mass of 5.4 kton. MINOS will improve our knowledge on $\Delta m^2$ significantly. Some improvement on the mixing angle measurements could also be expected. The experiment is expected to start in early 2005.

In Europe, a project called CNGS (CERN Neutrino to Gran Sasso) is in progress. 400 GeV proton beam from SPS at CERN produces high energy neutrinos whose mean energy is about 20–30 GeV. Neutrinos produced by this beam will be detected by detectors at Gran Sasso, which is 730 km away from the neutrino production point. OPERA is an experiment for the CNGS project.[20] This experiment is aimed to study neutrino oscillations by looking at the appearance of $\nu_\tau$ in the beam. The OPERA main detector consists of a 1.8 kton lead-emulsion target. Since it is not possible to scan all the images recorded in the emulsion, there are electronic detectors after each emulsion module in order to locate the interaction point in the emulsion. Only a portion of the emulsion that are located by the electronic detectors will be scanned to search for $\tau$ decay kinks. After 5 years of operation with $5 \times 10^{19}$ p.o.t. per year, the expected number of identified CC $\nu_\tau$ events are 4.3, 10.1 and 26.3 for $\Delta m^2 = 0.0016$, 0.0025, 0.004 eV$^2$, while the expected background is 0.65. The experiment will start in 2006.

The other experiment for the CNGS project will be ICARUS.[21] It is a liquid argon TPC detector. The total mass will be 3 ktons. The excellent imaging capability of the detector should make it possible to carry out various neutrino oscillation studies. 600 ton ICARUS detector has been tested on the ground successfully.[22] The same detector will be installed at the Gran Sasso Laboratory soon. The sensitivity for $\nu_\tau$ appearance is very similar to that of OPERA. It has also a high sensitivity for the $\theta_{13}$ measurement.

So far, we have discussed experiments that are in operation or under construction. There are as-yet-unobserved quantities related to neutrino oscillations: $\theta_{13}$, the sign of $\Delta m^2$ and the CP phase in the neutrino sector. These questions can be addressed by the next generation neutrino oscillation experiments. One possible experiment is the proposed JPARC-Kamioka neutrino experiment.[23] JPARC is a high intensity proton accelerator complex that is under construction at JAERI, Tokai, Japan. This experiment will use a very high intensity, low energy ($E_\nu < 1$ GeV) neutrino beam. The baseline length is 295 km. One of the main purposes of this experiment is the discovery of non-zero $\sin^2 \theta_{13}$. If the true $\sin^2 \theta_{13}$ is within 1/20 of the present reactor limit,[16] this experiment should be able to detect non-zero $\theta_{13}$.

It is also possible to study $\theta_{13}$ using anti electron neutrinos from nuclear power reactors. Unlike accelerator experiments, the disappearance of $\bar{\nu}_e$ is almost a pure measurement of $\theta_{13}$. Detailed studies have shown that it must be possible to improve the $\sin^2 \theta_{13}$ limit by a factor of 5 to 10.

There are now intense studies of designing experiments that detect CP violation effect. CP violation in the neutrino sector is considered to be the key to understand the baryon asymmetry of the Universe,[24] and therefore is very important. The CP violating phase is currently unknown and is expected to be observable if $\Delta m^2_{12}$ and $\Delta m^2_{23}$ are close enough so that both come into play in a single measurement and if the Jarlskog factor[25] $J = c_{12}c_{13}^2 c_{23}s_{12}s_{13}s_{23}(\sin \delta)$ is large enough. Recent solar neutrino and KamLAND data confirm that $\theta_{12}$ and $\Delta m^2_{12}$ are large enough.

One example for such experiments is the second stage of the JPARC-Kamioka project. The CP violation phase can be measured by observing the difference in the neutrino oscillation probabilities between $\nu_\mu \to \nu_e$ and $\bar{\nu}_\mu \to \bar{\nu}_e$. To observe this effect, a huge detector

(1 Mton Hyper-Kamiokande detector), a very high intensity proton accelerator (4 MW beam power) and about 8 years of operation are required.[23]

Even more ambitious projects are under serious discussion and R&D. These discussions assume a very high intensity neutrino beam produced by muon storage rings (neutrino factories), see for example Ref. 26. The basic idea is to build a muon storage ring with a long straight section where muons decay to produce a collimated neutrino beam. The muon decay is well understood and yields a very well defined beam spectrum. A $\mu^+$ decays to $e^+$, $\bar{\nu}_\mu$ and $\nu_e$. Therefore, if a $\mu^-$ is observed in a neutrino interaction, this event must be an oscillation signal (or a background due to an imperfect detector resolution). Therefore, experiments with muon storage rings are expected to be very sensitive to a very small $\theta_{13}$ value. In addition, since the storage of positive and negative muons are possible, it is possible to study the CP violation effect. Detailed studies have shown that the sensitivity of these experiments to the CP violation effect is high.

## 4. Summary

Data from various neutrino oscillation experiments to date already give fairly detailed information on the neutrino masses and mixing angles. Two mixing angles, $\theta_{12}$ and $\theta_{23}$ are large. $\Delta m_{23}^2$ is significantly larger than $\Delta m_{12}^2$. However, our understanding of the neutrino masses and mixing angles is not complete. We do not know how small $\theta_{13}$ is. We have no knowledge about the CP phase in the neutrino sector. In addition there are many open questions. Future long baseline experiments should address many of these important questions. It is likely that future neutrino oscillation experiments will continue to contribute to our understanding of the particle physics.

This work has been supported by the Japanese Ministry of Education, Science, Culture and Sports.

# References

1. T. Yanagida, in Proc. of the Workshop on the Unified Theory and Baryon Number in the Universe (KEK Report No. 79-18) p. 95 (1979).
2. M. Gell-Mann, P. Ramond, R. Slansky, in Supergravity, p. 315 (1979).
3. Q. R. Ahmad *et al.*, *Phys. Rev. Lett.* **89**, 011302 (2002).
4. S. Fukuda *et al.*, *Phys. Lett. B* **539**, 179 (2002).
5. K. Eguchi *et al.*, *Phys. Rev. Lett.* **90**, 021802 (2003).
6. P. Cunha de Halanda and A. Yu. Smirnov, *JCAP* **0302**, 001 (2003).
7. M. B. Smy *et al.*, hep-ex/0309011.
8. S. N. Ahmed *et al.*, nucl-ex/0309004.
9. Y. Fukuda *et al.*, *Phys. Rev. Lett.* **81**, 1562 (1998).
10. Y. Fukuda *et al.*, *Phys. Lett. B* **335**, 237 (1994).
11. S. Hatakeyama *et al.*, *Phys. Rev. Lett.* **81**, 2016 (1998).
12. M. Sanchez *et al.*, hep-ex/0307069
13. M. Ambrosio *et al.*, *Phys. Lett. B* **566**, 35 (2003).
14. C. Saji, for the Super-Kamiokande collaboration, in Proc. of the 28th International Cosmic Ray Conference, Tsukuba, Japan, July-Aug. 2003, Vol. 3, p. 1267.
15. M. H. Ahn *et al.*, *Phys. Rev. Lett.* **90**, 041801 (2003).
16. M. Apollonio *et al.*, *Phys. Lett. B* **466**, 415 (1999).
17. K. Nishikawa, talk presented at the XXI International Symposium on Lepton and Photon Interactions at High Energies, Fermilab, USA, Aug. 2003.
18. G. L. Fogli *et al.*, hep-ph/0308055.
19. D. Michael, *Nucl. Phys. B (Proc. Suppl.)* **118**, 189 (2003).
20. OPERA collaboration, M. Guler, *et al.*, CERN-SPSC-P-318, LNGS-P25-00 (2000).
21. A. Rubbia, *Nucl. Phys. B (Proc. Suppl.)* **91**, 223 (2001).
22. O. Palamara, for the ICARUS Collaboration, *Nucl. Phys. B (Proc. Suppl.)* **110**, 329 (2002).
23. Y. Itow *et al.*, hep-ex/0106019.
24. M. Fukugita and T. Yanagida, *Phys. Lett. B* **174**, 45 (1986).
25. C. Jariskog, *Phys. Rev. Lett.* **55**, 1039 (1985).
26. S. Geer, *J. Phys. G: Nucl. Part. Phys.* **29**, 1485 (2003).

# Chapter 7

# Solar and Atmospheric Neutrinos*

Takaaki Kajita

*Research Center for Cosmic Neutrinos, Institute for Cosmic Ray Research, Univ. of Tokyo, Kashiwa-no-ha 5-1-5, Kashiwa, Chiba 277-8582, Japan*
*kajita@icrr.u-tokyo.ac.jp*

RECENT data on solar and atmospheric neutrinos and their constraints on the neutrino mass and mixing are presented. Especially, the results from a 50 kton water Cherenkov detector, Super-Kamiokande, are discussed in detail.

## 1. Solar Neutrinos

Five solar neutrino experiments[1–5] confirmed the basic mechanism of the energy generation in the Sun. However, the observed fluxes from these experiments were lower than the Standard Solar Model (SSM) predictions.[6] Neutrino oscillations[7] can explain the existing solar neutrino data. For the definite confirmation of solar neutrino oscillations, experimental results which cannot be predicted by any solar models are highly desirable. Furthermore, the parameter region of neutrino oscillations should be uniqly determined. It is predicted that the spectrum of the $^8$B solar neutrinos should be distorted significantly for the cases of the "small angle MSW

---

*This article was originally published in *Particles, Strings and Cosmology*, pp. 386–393 (2000).

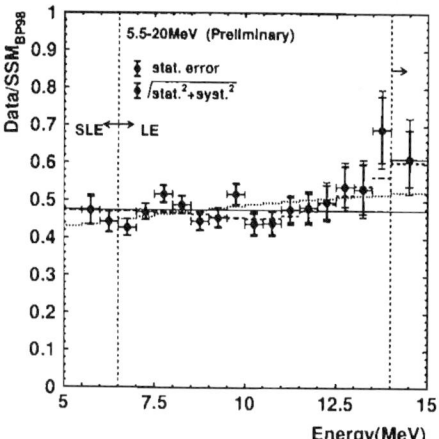

Figure 1.   Plot of Data/SSM as a function of electron energy. The dotted and dashed histograms show the expected energy spectrum for a typical small-mixing solution ($\sin^2 2\theta = 5 \times 10^{-3}$, $\Delta m^2 = 5 \times 10^{-6}$ eV$^2$) and the best-fit vacuum-oscillation solution ($\sin^2 2\theta = 0.79$, $\Delta m^2 = 4.3 \times 10^{-10}$ eV$^2$), respectively.

solution" and the "vacuum oscillation solution", or the day-night effect should be observed for a region of the "large angle MSW solution".

Super-Kamiokande detects $^8$B solar neutrinos through $ve \rightarrow ve$. The measured $^8$B solar $v$ flux above 6.5 MeV during 825 days of the observation time was $0.475 \, ^{+0.008}_{-0.007}$(stat.) $\pm 0.013$(sys. of the data) of the SSM prediction. Solar neutrino events in a lower energy region (5.5 to 6.5 MeV) were also observed (524 days of data). The observed flux in this energy range was consistent with the higher energy data, see Figure 1.

Super-Kamiokande provides the information for both the energy spectrum and the day-night effect. The day-night data were expressed as: $Night/Day = 1.067 \pm 0.033$(stat.) $\pm 0.013$(sys.). The possible excess of the flux in the night time was $2\sigma$ level and was not significant yet.

Figure 1 shows the energy spectrum of the recoil-electrons divided by the SSM prediction. The Data/SSM values in the endpoint region were higher than the average. The statistical significance of the possible distortion of the energy spectrum was tested by a $\chi^2$ method. The absolute flux value was assumed to be a free parameter and only the shape of the distributions of the data and the expectations was compared. The $\chi^2$ values were obtained to be 24.3, 25.0 and 17.4 (DOF=17) for the no distortion case (i.e., no oscillation effect), the small-mixing MSW case and the best fit in the vacuum oscillation solution, respectively. The energy spectrum measurement was not conclusive. More data are needed for further understanding of the solar neutrino problem.

Recently SNO heavy water experiment started taking data. Data from SNO on $v_e D \rightarrow e^- pp$ reactions (CC) are expected to be presented in 2000. It is expected that we may get important information on neutrino oscillations by comparing the Super-Kamiokande data on $ve^- \rightarrow ve^-$ and the SNO CC data.[8] The SNO CC data only contain $v_e$ CC events. On the other hand, $ve^- \rightarrow ve^-$ measurement is also sensitive to $v_\mu$ and $v_\tau$ with the reduced cross section of $1/(6$ to $7)$ relative to that of $v_e$. Therefore, (assuming the SSM flux) if $x\%$ of the solar $v_e$'s are oscillating to $v_\mu$'s or $v_\tau$'s, the observed signal by SNO must be $(1 - x)$ of the SSM prediction. The expected signal by the Super-Kamiokande is $(1 - x) + 1/(6$ to $7) \cdot x$ of the SSM. By taking $(Data/SSM)_{SNO}/(Data/SSM)_{SK}$ we should be able to see the neutrino oscillation effect independent of the SSM prediction.

## 2. Atmospheric Neutrinos

Cosmic ray interactions in the atmosphere produce neutrinos. Kamiokande reported that the measured values of the $(\mu/e)$ ratio were significantly smaller than expected.[9,10] Also a zenith-angle dependent deficit of $\mu$-like events was observed by Kamiokande[10] at high energies. These observations strongly suggested neurino oscillations, and therefore Kamiokande estimated the allowed parameter regions of neutrino oscillations. However, due to the relatively small statistics, both $v_\mu \leftrightarrow v_e$ and $v_\mu \leftrightarrow v_\tau$ oscillations were allowed. Recently, a long baseline reactor experiment, CHOOZ,[13] excluded the $\nu_\mu \leftrightarrow \nu_e$ solution of the atmospheric neutrino problem. In 1998, Super-Kamiokande reported that the atmospheric neutrino data gave evidence for neutrino oscillations.[14] After that, the statistics of the Super-Kamiokande data were increased by about a factor of two: a total of 8145 fully-contained (FC) events and 563 partially-contained (PC) events were observed in a 61 kton·year exposure.

For the analysis of FC events, only single-ring events were used. Single-ring events were identified as $e$-like or $\mu$-like. The FC events were separated into "sub-GeV" ($E_{vis} < 1330$ MeV) and "multi-GeV" ($E_{vis} > 1330$ MeV) samples. All the PC data were categorized as multi-GeV $\mu$-like.

Figure 2 summarizes the $(\mu/e)_{data}/(\mu/e)_{MC}$ measurements. The observed values of this ratio by recent experiments (Super-Kamiokande and Soudan-2[12]) were significantly smaller than unity and were consistent with the old Kamiokande (and the IMB sub-GeV) results.

The zenith angle distributions from Super-Kamiokande are shown in Figure 3. The $\mu$-like data exhibited a strong up-down asymmetry in zenith angle ($\Theta$) while no significant asymmetry was observed in the $e$-like data.

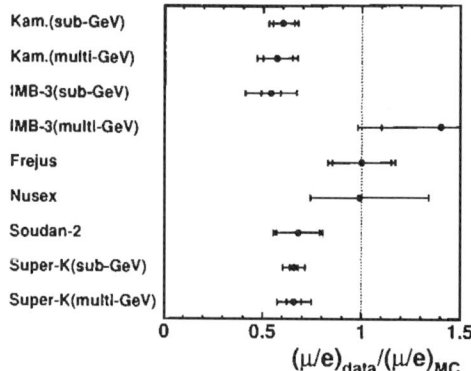

Figure 2.   Summary of the atmospheric $(\mu/e)_{data}/(\mu/e)_{MC}$ ratio measurement by various experiments. The inner error bars show the statistical errors and the outer error bars show the total errors.

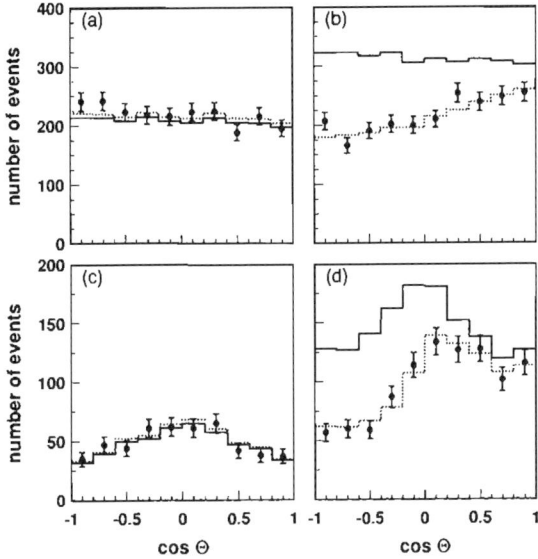

Figure 3.   Zenith angle distributions for (a) sub-GeV *e*-like, (b) sub-GeV *v*-like, (c) multi-GeV *e*-like and (d) multi-GeV (FC+PC) *μ*-like eventgs observed in Super-Kamiokande. CosΘ = 1 means down-going particles. The solid histograms show the MC prediction without neutrino oscillations. The dashed histograms show the MC prediction for $v_\mu \leftrightarrow v_\tau$ oscillations with $\sin^2 2\theta = 1$ and $\Delta m^2 = 2.8 \times 10^{-3}$ eV$^2$. In the oscillation histograms, the absolute normalization was adjusted to get the minimum $\chi^2$.

Energetic atmospheric $v_\mu$'s passing through the Earth interact with rock surrounding the detector and produce muons via CC interactions. These neutrino events are observed as upward going muons. Super-Kamiokande[15] observed 1196 (285) upward through-going (stopping) muon events (including 8.6 ± 0.8 (20.2 ± 8.0) estimated background events) during 1074 (1053) detector live days. Figure 4 shows the zenith-angle distributions of the upward-going muon fluxes. The prediction for the through-going muons had a flatter zenith-angle distribution than the data. Similar data were obtained in Kamiokande[16] and MACRO.[17] The observed flux of upward stopping muons by Super-Kamiokande was significantly smaller than the prediction. These data are explained by neutrino oscillations.

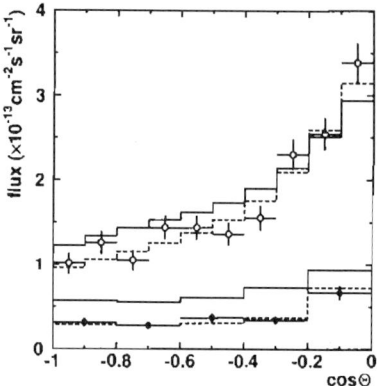

Figure 4. Zenith angle distribution of upward-going muon flux observed in Super-Kamiokande. Blank (filled) circles with error bars show the upward through-going (stopping) muon data the histograms show the corresponding predictions. Error bars show statistical + experimental systematic errors. Estimated backgrounds are subtracted. The solid histograms show the expected fluxes for the null neutrino oscillation case. The dashed histograms show the expected fluxes for the $\nu_\mu \leftrightarrow \nu_\tau$ oscillation case with $\sin^2 2\theta = 1.0$ and $\Delta m^2 = 2.8 times 10^{-3}$ eV$^2$.

Since the contained events and upward-going muon events consistently suggested the neutrino oscillations, an allowed region was obtained by using all the atmospheric neutrino data from Super-Kamiokande.[18] The allowed region is shown in Figure 5 together with the allowed regions from Kamiokande,[16] Soudan-2[12] and MACRO.[17] The best fit point was found at ($2.8 \times 10^{-3}$ eV$^2$, 1.0) (including the unphysical region of $\sin^2 2\theta > 1$). The 90% C.L. allowed region of the oscillation parameters was; $2 \times 10^{-3} < \Delta m^2 < 5 \times 10^{-3}$ eV$^2$ and $\sin^2 2\theta > 0.88$.

As an extension of this analysis, Super-Kamiokande carried out a 3 flavor neutrino oscillation analysis. It was assumed that the mass difference between the lightest and the second lightest neutrinos is too small and therefore the effect of the oscillation is invisible in atmospheric neutrinos. The allowed region was obtained on the 3 dimensional space of $\sin^2 \theta_{13}$, $\sin^2 \theta_{23}$ and $\Delta m^2 (\equiv \Delta m^2_{13} = \Delta m^2_{23})$. An allowed region on the $\sin^2 \theta_{13}$ and $\sin^2 \theta_{23}$ plane is shown in Figure 5. (Note that the axes show $\sin^2 \theta_{ij}$, not $\sin^2 2\theta_{ij}$.) Super-

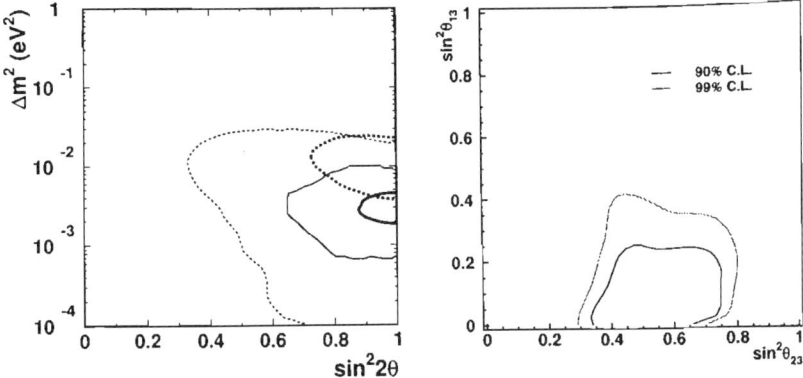

Figure 5. (Left) 90% C.L. allowed parameter region of $\nu_\mu \leftrightarrow \nu_\tau$ neutrino oscillations by a combined analysis of FC, PC, upward stopping muon and upward through-going muon events from Super-Kamiokande (thick solid line, preliminary). Results from Kamiokande (thick dashed line), Soudan-2 (thin dashed line) and MACRO (thin solid line) are also shown. (Right) Allowed region on a $\sin^2 \theta_{13}$ and $\sin^2 \theta_{23}$ plane obtained from the Super-Kamiokande FC+PC data.

Kamiokande found no evidence for non-zero $\sin^2 \theta_{13}$. At present, the constraint from CHOOZ[13] on $\sin^2 \theta_{13}$ in the $\Delta m^2$ range between 2 and $5 \times 10^{-3}$ eV$^2$ is much stronger than that from Super-Kamiokande.

Super-Kamiokande also studied if the energy dependence of the neutrino oscillation is really as expected by the standard neutrino oscillations generated by neutrino mass and mixing. For this purpose, the atmospheric neutrino data (including upward going muons) were fitted with the form of $P(\nu_\mu \to \nu_\mu) = 1 - \sin^2 \alpha \cdot \sin^2(\beta \cdot L \cdot E^n)$, where $\alpha, \beta$ and $n$ were fitted parameters.[19] The energy dependence expected by the standard neutrino oscillation generated by neutrino mass and mixing ($n = -1$) was favored by the data (the fitted value of $n$ was about $-1.0 \pm 0.1$), and models which predict other energy dependences of the oscillations were strongly disfavored.

The contained event data could also be explained by the $\nu_\mu \leftrightarrow \nu_{sterile}$ neutrino oscillations. Where $\nu_{sterile}$ is a hypothetical neutrino-like particle which does not interact with matter.

Super-Kamiokande studied zenith angle distribution of NC enriched sample. Since a $\nu_{sterile}$ does not interact with matter via NC,
there should be a deficit of upward going NC events for $\nu_{\mu} \leftrightarrow \nu_{sterile}$.
In addition, it is possible to discriminate these two possibilities by
using the matter effect.[20] In the case of $\nu_{\mu} \leftrightarrow \nu_{\tau}$ oscillations, the matter effect does not change the oscillation probability, however in the
case of $\nu_{\mu} \leftrightarrow \nu_{sterile}$ oscillations, matter effect may change the oscillation probability significantly. The difference of the neutrino oscilla-
tion probability is large only for high energy atmospheric neutrinos
traveling through the Earth.

NC enriched sample was made by; multi-ring events, most energetic ring must be $e$-like and $E_{vis} > 400$ MeV. The fraction of NC
events in this sample is estimated to be 29% for no neutrino oscillations. Figure 6(a) shows the zenith angle distribution of these events
together with predictions of neutrino oscillations. An up-down ratio, $Up(-1 < \cos\Theta < -0.4)/Down(0.4 < \cos\Theta < 1)$, of the data
was $0.99 \pm 0.07 \pm 0.005$ while the predictions at $(\sin^2 2\theta = 1, \Delta m^2 = 3 \times 10^{-3})$ were $0.93 \pm 0.03$(sys.) for $\nu_{\mu} \leftrightarrow \nu_{\tau}$ and $0.81 \pm 0.02$(sys.) for

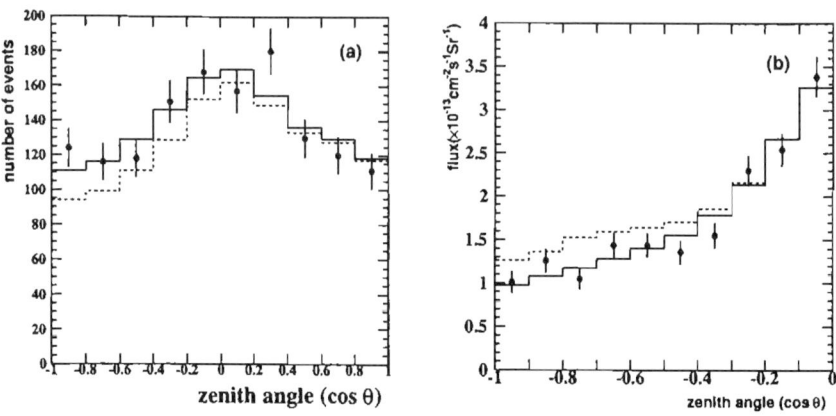

Figure 6.    Zenith angle distributions for; (a) NC enriched multi-ring events and (b) upward
through-going muon events observed in Super-Kamiokande. In (a) and (b), solid (dashed)
histograms show the expectations for $\nu_{\mu} \leftrightarrow \nu_{\tau}$ ($\nu_{\mu} \leftrightarrow \nu_{sterile}$). In these figures, $\sin^2 2\theta = 1$ is
assumed for the expectations.

$\nu_\mu \leftrightarrow \nu_{sterile}$. The expected up-down ratio for $\nu_\mu \leftrightarrow \nu_{sterile}$ neutrino oscillation hypothesis did not agree with the data at $2\sigma$ level.

To separate the two neutrino oscillation hypotheses using the matter effect, Super-Kamiokande studied the upward through-going muons and high-energy PC events. Figure 6(b) shows the zenith angle distribution of the upward through-going muons together with predictions of neutrino oscillations. A vertical-horizontal ratio, $Vertical(-1 < \cos\Theta < -0.4)/Horizontal(-0.4 < \cos\Theta < 0)$, of the data was $0.77 \pm 0.045 \pm 0.007$ while the predictions at ($\sin^2 2\theta = 1$, $\Delta m^2 = 3 \times 10^{-3}$) were $0.76 \pm 0.024$(sys.) for $\nu_\mu \leftrightarrow \nu_\tau$ and $0.91 \pm 0.029$(sys.) for $\nu_\mu \leftrightarrow \nu_{sterile}$. The expected vertical-horizontal ratio for $\nu_\mu \leftrightarrow \nu_{sterile}$ neutrino oscillation hypothesis did not agree with the data at $2.5\sigma$ level. A similar (but weaker) result was obtained by the PC events.

Finally, the two oscillation hypotheses were tested by combining these studies by a $\chi^2$ method. Most of the parameter regions suggested by the analysis of FC events for $\nu_\mu \leftrightarrow \nu_{sterile}$ were disfavored at about 99% C.L. (preliminary). On the other hand, $\nu_\mu \leftrightarrow \nu_\tau$ oscillation hypothesis did not contradict with the data in this analysis.

## 3. Summary

Both the zenith angle distribution of $\mu$-like events and the ($\mu/e$) values from Super-Kamiokande were significantly different from the predictions in the absence of neutrino oscillations. The data were in good agreement with $\nu_\mu \leftrightarrow \nu_\tau$ oscillations. This conclusion was supported by the upward-going muon results, and was consistent with the results from Kamiokande, Soudan-2 and MACRO. The 90% C.L. allowed region of neutrino oscillation parameters from Super-Kamiokande was: $2 \times 10^{-3} < \Delta m^2 < 5 \times 10^{-3}$ eV$^2$ and $\sin^2 2\theta > 0.88$. The matter effect and the NC events were used to discriminate $\nu_\mu \leftrightarrow \nu_\tau$ and $\nu_\mu \leftrightarrow \nu_{sterile}$ oscillations. The preliminary result disfavored the $\nu_\mu \leftrightarrow \nu_{sterile}$ oscillation hypothesis at about 90% C.L.

To distinguish the solutions of the solar neutrino problem, the day-night effect and the energy spectrum of $^8$B solar neutrinos were studied. The present Super-Kamiokande data did not distinguish these solutions. The comparison of the Super-Kamiokande and the SNO data may give another important information on solar neutrino oscillations in the near future.

This work was partly supported by the Japanese Ministry of Education, Science, Sports and Culture.

## References

1. B. T. Cleveland *et al.*, *Ap. J.* **496**, 505 (1998).
2. Y. Fukuda *et al.*, *Phys. Rev. Lett.* **77**, 1683 (1996).
3. J. N. Abdurashitov *et al.*, *Phys. Rev. Lett.* **83**, 4686 (1999).
4. W. Hampel *et al.*, *Phys. Lett. B* **447**, 127 (1999).
5. Y. Fukuda *et al.*, *Phys. Rev. Lett.* **81**, 1158 (1998); *Phys. Rev. Lett.* **82**, 1810 (1999); *Phys. Rev. Lett.* **82**, 2430 (1999).
6. J. N. Bahcall, S. Basu and M. H. Pinsonneault, *Phys. Lett. B* **433**, 1 (1998).
7. L. Wolfenstein, *Phys. Rev. D* **17**, 2369 (1978); S. P. Mikheyev and A. Yu. Smirnov, *Sov. J. Nucl. Phys.*, **42**, 1441 (1985); S. P. Mikheyev and A. Yu. Smirnov, *Nuovo Cimento C* **9**, 17 (1986).
8. J. N. Bahcall, P. I. Krastev and A. Yu. Smirnov, *Phys. Lett. B* **477**, 401 (2000).
9. K. S. Hirata *et al.*, *Phys. Lett. B* **205**, 416 (1988); *Phys. Lett. B* **280**, 146 (1992).
10. Y. Fukuda *et al.*, *Phys. Lett. B* **335**, 237 (1994).
11. D. Casper *et al.*, *Phys. Rev. Lett.* **66**, 2561 (1991); R. Becker-Szendy *et al.*, *Phys. Rev. D* **46**, 3720 (1992); R. Clark *et al.*, *Phys. Rev. Lett.* **79**, 345 (1997).
12. W. W. M. Allison *et al.*, *Phys. Lett. B* **391**, 491 (1997); *Phys. Lett. B* **449**, 137 (1999); W. A. Mann, hep-ex/9912007.
13. M. Apollonio *et al.*, *Phys. Lett. B* **420**, 397 (1998); *Phys. Lett. B* **466**, 415 (1999).
14. Y. Fukuda *et al.*, *Phys. Lett. B* **433**, 9 (1998); *Phys. Lett. B* **436**, 33 (1998); *Phys. Rev. Lett.* **81**, 1562 (1998).
15. Y. Fukuda *et al.*, *Phys. Rev. Lett.* **82**, 2644 (1999); Y. Fukuda *et al.*, *Phys. Lett. B* **467**, 185 (1999).
16. S. Hatakeyama *et al.*, *Phys. Rev. Lett.* **81**, 2016 (1998).
17. M. Ambrosio *et al.*, *Phys. Lett. B* **434**, 451 (1998); M. Spurio, for the MACRO collaboration, hep-ex/9908066.
18. The Super-Kamiokande collaboration, draft in preparation.
19. G. L. Fogli *et al.*, *Phys. Rev. D* **60**, 053006 (1999).
20. Q. Y. Liu, S. P. Mikheyev and A. Yu. Smirnov, *Phys. Lett. B* **440**, 319 (1998); P. Lipari and M. Lusignoli, *Phys. Rev. D* **58**, 073005 (1998).